Studies in Logic
Volume 16

Foundations of the Formal Sciences VI

Probabilistic Reasoning and Reasoning with Probabilities

Studies in Logic Series Editor
Dov Gabbay   dov.gabbay@kcl.ac.uk

Foundations of the Formal Sciences VI

Probabilistic Reasoning and Reasoning with Probabilities

Edited by

Benedikt Löwe
Eric Pacuit
and
Jan-Willem Romeijn

ISBN 978-1-904987-15-4

College Publications
Scientific Director: Dov Gabbay
Managing Director: Jane Spurr
Department of Computer Science
King's College London, Strand, London WC2R 2LS, UK

http://www.collegepublications.co.uk

Original cover design by Richard Fraser
Created by orchid creative   www.orchidcreative.co.uk
Printed by Lightning Source, Milton Keynes, UK

Table of Contents

Historical Papers

Preface

Probabilistic methods have become an important tool in a variety of disciplines. This includes computer science (probabilistic computation and automata, randomness, Bayesian networks), mathematics (probabilistic proofs), artificial intelligence (reasoning under uncertainty), epistemology (Bayesian epistemology) and linguistics (probabilistic grammars). And of course, from the beginning probabilistic methods have been heavily used in decision theory and game theory. In most disciplines concerned with rationality and agency, probability theory plays an important role.

The long-standing philosophical debate over the precise meaning of probabilistic statements is often separate from the discussion on applications of probabilistic methods. This debate raises a number of issues that are crucial to the interpretation and application of results achieved by probabilistic methods. The sixth conference on the Foundations of the Formal Sciences was an interdisciplinary forum for researchers that use probabilistic methods in their respective fields, and researchers that are concerned with the philosophical interpretation of probability. The goal of the meeting was to exchange ideas, approaches and techniques, to facilitate discussions about the applicability of probabilistic methods, and to help ground the foundational debates in the concerns of the users of probabilistic methods.

The FotFS conference series.

The conference was the sixth in a conference series on the "Foundations of the Formal Sciences" (FotFS). This is a series of interdisciplinary conferences in mathematics, philosophy, computer science and linguistics. The main goal is to reestablish the traditionally strong links between these areas of research, some of which have been lost in the past decades. FotFS started in 1999 as a small German workshop in Berlin. Its defining features were present from the very first meeting onwards: a strong interdisciplinary spirit, a focus on technical talks that nevertheless reach out to researchers from other communities, and a (non-exclusive) focus on young researchers.

After its inaugural meeting in Berlin, FotFS was funded as a "PhD Euro-Conference" by the European Community, the DFG (Deutsche Forschungsgemeinschaft) and the BIGS (Bonn International Graduate School). Each of the meetings has a distinctive topic specifying some part of the foundations of formal sciences to be investigated in an interdisciplinary way. FotFS II dealt with Applications of Mathematical Logic in Philosophy and Linguistics, and FotFS III with "Complexity in Mathematics and Computer Science". FotFS IV was a meta-conference discussing the topic of the series under the header "The History of the Concept of the Formal Sciences". FotFS V concerned the topic "Infinite Games". By now, the seventh conference (FotFS VII) has been held in Brussels in October 2008, under the title "Bringing together philosophy and sociology of science".

Foundations of the Formal Sciences VI (FotFS VI) was held on the topic "Reasoning about Probabilities and Probabilistic Reasoning" from the 2nd to 5th of May 2007 in Amsterdam. The members of the Scientific and Organizing Committee were Horacio Arló-Costa, Benedikt Löwe, David Makinson, Eric Pacuit, and Jan-Willem Romeijn. Since the topic of the conference overlapped considerably with the topic of the Leverhulme Trust network **progicnet**, there was a special session where members of this network presented their work. In the following, we list all of the presentations given at the conference, ordered alphabetically by author names.

David Atkinson (Groningen, The Netherlands): *A Case Where Chance and Credence Coincide*

Conor Barry (Paris, France): *Certainly Probable: Belief and Aleatory Probability in Hume's Theory of Induction*

Luc Bovens (London, United Kingdom): *Dutch Books, Group Decision-Making, the Tragedy of the Commons and Strategic Jury Voting* (INVITED TALK)

Timothy Childers (Prague, Czech Republic): *Non-pragmatic arguments for Bayesianism*

David Corfield (Canterbury, United Kingdom): *What's Happening in Machine Learning Today?* (INVITED TALK)

Helen De Cruz (Brussels, Belgium): *The role of intuitive probabilistic reasoning in scientific theory formation*

Carla A.D.M. Delgado (Rio de Janeiro, Brazil) and Mario Benevides (Rio de Janeiro, Brazil): *Model Checking Knowledge in Probabilistic Systems*

Isabelle Drouet (Paris, France): *Can there be a propensity interpretation of conditional probabilities?*

Branden Fitelson (Berkeley CA, United States of America): *Epistemological Critiques of "Classical" Logic: Two Case Studies* (INVITED TALK)

Maria Carla Galavotti (Bologna, Italy): *Probability: one or many?* (INVITED TALK)

Anne-Sophie Godfroy-Genin (Paris, France): *From the doctrine of probability to the theory of probabilities: the emergence of modern probability calculus* (INVITED TALK)

Norma B. Goethe (Cordoba, Argentina): *The fundamental shift in the metaphor through which philosophers conceive of learning*

Peter Grünwald (Amsterdam, The Netherlands): *Statistics without Stochastics* (INVITED TALK)

Joseph Halpern (Ithaca NY, United States of America): *Redoing the Foundations of Decision Theory* (INVITED TALK)

Joel David Hamkins (New York NY, United States of America; Amsterdam, The Netherlands): *The Halting Problem is decidable on a set of asymptotic probability one*

Jeffrey Helzner (New York NY, United States of America): *Indeterminacy in the Combining of Attributes*

Brian Hill (Paris, France): *Beliefs: identification and change*

Barteld Kooi (Groningen, The Netherlands): *Dynamic Update with Probabilities*

Theo Kuipers (Groningen, The Netherlands): *Empirical Progress and Truth Approximation by the "Hypothetical Probabilistic (HP-)method"*

Jürgen Landes (Manchester, United Kingdom), Jeff Paris (Manchester, United Kingdom) and Alena Vencovska (Manchester, United Kingdom): *The Principle of Conformity and Spectrum Exchangeability*

Bert Leuridan (Ghent, Belgium): *Combining probability and logic: Embedding causal discovery in a logical framework*

Ondrej Majer (Prague, Czech Republic): *Fuzzy logic and betting games*

Klaus Nehring (Davis CA, United States of America): *Like-Minded Agents as a Foundation for Common Priors*

Martin Neumann (Osnabrück, Germany): *Measuring the Uncertain*

Kenneth Presting (Chapel Hill NC, United States of America): *Probability Spaces for First-Order Logic*

Jeanne Peijnenburg (Groningen, The Netherlands): *Probabilistic Justification and the Regress Problem*

Paolo Rocchi (Roma, Italy): *Eclectic Interpretation of the Probability: A Question of Convenience?*

Jonah Schupbach (Pittsburgh PA, United States of America): *On the Alleged Impossibility of Bayesian Coherentism*

Teddy Seidenfeld (Pittsburgh PA, United States of America): *Concepts of Independence for Full Conditional Measures and Sets of Full Conditional Measures* (INVITED TALK)

Jan Sprenger (Bonn, Germany; London, United Kingdom): *Surprise and Evidence in Model Checking*

There were three more talks accepted by the scientific committee but not presented at the conference:

Aidan Lyon (Canberra, Australia): *Probability in Evolutionary Theory*

Weng Hong Tang (Canberra, Australia): *Rationality Constraints on Credences*

Emil Weydert (Luxembourg, Luxembourg): *Quality, quantity, and beyond. On nonmonotonic probabilistic reasoning*

The conference FotFS VI was financially supported by the *Nederlandse Organisatie voor Wetenschappelijk Onderzoek* (NWO; DN 613.080.000 CN 2006/09615/EW), the Marie Curie Research Training Site GLoRiClass (MEST-CT-2005-020841), the Leverhulme Trust network progicnet, and the *Institute for Logic, Language and Computation* (ILLC) of the Universiteit van Amsterdam. We would like to thank our volunteer helpers who dealt with practical matters during the conference and without whose help the conference couldn't have been held: Cédric Dégremont, Sujata Ghosh, Marc Staudacher, Joel Uckelman, Sara Uckelman, Marjan Veldhuisen, and Jonathan Zvesper. The typesetting and printing of this volume was funded by Eric Pacuit's VIDI project *"A formal analysis of social procedures"* (016.094.345) and the Marie Curie Research Training Site GLoRiClass (MEST-CT-2005-020841), and we acknowledge the bibliographical help of Edgar Andrade.

This volume.

This volume is the proceedings volume of FotFS VI for which we received twenty submissions. Thirty expert referees helped us to select the twelve papers published in this volume, falling into three overlapping approaches to probability: formal, philosophical, and historical. In the following, we briefly introduce the contributions, their respective topics, and some of the ways in which they relate.

The first cluster of contributions concerns formal work in statistics and decision theory. *Cozman and Seidenfeld* present cutting edge research on two important technical advances in representing uncertainty, to wit, the use of network structures like Bayesian networks, and the formulation of probability theory in terms of full conditional measures, sometimes referred to as Popper functions. It is shown that standard notions of independence and irrelevance, as used in the theory of Bayesian networks for ordinary probability functions, cannot be applied straightforwardly when using conditional measures. The paper by *Helzner* also concerns alternatives to standard probability theory as an expression of uncertainty. Specifically, Helzner considers Levi's notion of credal states. However, the focus of his paper is different from the contribution by Cozman and Seidenfeld. Helzner connects credal sets to concepts from decision theory, in particular to choice functions, and shows how such choice functions can be used to characterise and develop a qualitative version of Levi's decision theory.

The contribution by the research collective **progicnet**, consisting of *Haenni, Romeijn, Wheeler* and *Williamson*, is connected to both the aforementioned papers. On the one hand it discusses the use of networks, and on the other it employs credal states. The paper by Progicnet illustrates a general framework for probabilistic logics, which employs so-called credal networks, in an application to statistical, and hence inductive, inference. The paper by *Landes, Paris, and Vencovská*, finally, is naturally related to this latter contribution, since it also concerns inductive inference, i.e., inference running from data to predictions. As part of their programme to revitalise Carnapian inductive logic, they discuss some notions that are central to both inductive logic and statistics, primarily exchangeability. They develop a new version of this notion, which they call spectrum exchangeability, and they show how it hangs together with another natural concept within inductive logic, conformity.

The second cluster of papers concerns probability in the context of philosophical problems. Three of these concern the use of probability as an expression of degrees of belief. The contribution by *Bovens and Rabinowicz* confronts so-called Dutch book arguments, i.e., arguments towards adopting probability functions as the expression of rational degrees of belief, with collective or game-theoretic rationality. It is shown, first, that groups seem liable to Dutch books in ways that individuals are not, and second, that there are formal relations between such collective Dutch book situations and other well-known problems with collective rationality, specifically the tragedy of the commons. The contribution by *Childers* is also concerned with Dutch books. He discusses a number of philosophical interpretations of Dutch books, and compares these with an alternative due to DeGroot, leading to a reassessment of the use of Dutch books. *Peijnenburg*'s contribution, finally, is also concerned with probability as degree of belief, but it targets the idea that beliefs, expressed as probabilities, are necessarily based on certainty. A new argument is given to the effect that there need not be any such certainty to stop short a regress of uncertain opinions. The upshot is that nontrivial uncertain beliefs can be maintained even if they are only underpinned by other uncertainties.

Two further contributions fall within the philosophical cluster, more specifically within the philosophy of science. Both of them pertain to probability as an expression of objective chance. The paper by *De Cruz* is a case study of how probabilities as long-term frequencies are used in scientific inference. The key observation of the paper is that the intuitive use of these frequencies, which is often associated with mistakes in reasoning, can fulfill an important role in scientific creativity and understanding. The paper thereby contributes to an analysis of probability as frequency. The paper by *Drouet* concerns objective chance in another way, namely probability as propensity. Drouet discusses Humphrey's paradox, which concerns propensities and conditional probabilities. By philosophical argument it is shown that an interpretation of conditional probability as propensities is feasible, and Drouet ends by advancing such an interpretation herself.

Finally, this volume contains three contributions on the history of probability. The paper by *Galavotti* gives a broad overview of the history of the concept of probability. Particular attention is given to the question whether and in what way the subjective theory of probability, associated with Ramsey and de Finetti, can accommodate an objective notion of chance, as it is employed in science. She argues that the views of Ramsey and Jeffreys provide ample space for addressing objectivity within an epistemic view on probability. The contribution of *Godfroy-Genin* also concerns probability as a concept in between objective chance and degrees of belief. She traces the origins of probability theory back to the *Port Royal* logic of Arnaud and his contemporaries, and she shows that the development of this logic played a crucial role in transforming the probability theory of these days, which was primarily concerned with the quantitative analysis of objective probability in gambling, into a theory on epistemic uncertainty. *Neumann*, finally, focuses on objective chance rather than epistemic uncertainty. By an analysis of von Kries' use of Boltzmann's work in the development of the notion of *Spielraum* by the former, Neumann presents a way of understanding propensities that differs from those that are currently on offer, thereby bridging the gap between historical and systematic analysis.

Of course, there are numerous other natural connections between the contributions in this volume. Neumann's insights on propensities are strongly related to Drouet's proposal on conditional propensities. Peijnenburg's analysis ties in with the more mathematical treatment of conditional measures by Cozman and Seidenfeld. Childers paper picks up where the historical analysis of Galavotti on Ramsey, especially his use of Dutch book arguments, leaves off. The list goes on... We invite the reader to enjoy this volume and find all of its $\binom{12}{2}$ themes!

April 2009, Amsterdam, Stanford, and Groningen   B. L.   E. P.   J.-W. R.

The conference photo.

Benedikt **Löwe**, Eric **Pacuit**, Jan-Willem **Romeijn** (*eds.*)
Foundations of the Formal Sciences VI
Reasoning about Probabilities and Probabilistic Reasoning

Independence for Full Conditional Measures and their Graphoid Properties

Fabio G. Cozman[1], Teddy Seidenfeld[2]*

[1] Escola Politécnica da Universidade de Sao Paulo, Avenida Prof. Luciano Gualberto, travessa 3 nᵒ 380, CEP, 05508-970, São Paulo, Brazil

[2] Department of Philosophy, Carnegie Mellon University, Baker Hall 135, Pittsburgh PA 15213, United States of America

E-mail: fgcozman@usp.br,teddy@stat.cmu.edu

1 Introduction

In this paper we wish to consider independence concepts associated with *full conditional measures* [23]. That is, we wish to allow conditioning to be a primitive notion, defined even on *null events* (events of zero probability). The advantages of a probability theory that takes conditioning as a primitive have been explored by many authors, such as de Finetti [20] and his followers [12, 16], Rényi [47], Krauss [37] and Dubins [23]. A significant number of philosophers have argued for serious consideration of null events [1, 29, 38, 41, 45, 48, 49, 50, 51], as have economists and game theorists [7, 8, 31, 32, 33, 42]. Null events, or closely related concepts, have also appeared repeatedly in the literature on artificial intelligence: ranking and ordinal measures [17, 28, 49, 59] have direct interpretations as "layers" of full conditional measures [12, 26, 43]; some of the most general principles of default reasoning can be interpreted through various types of lexicographic probabilities [5, 6, 25, 36, 44]; and as a final example, the combination of probabilities and logical constraints in expert systems must often deal with zero probabilities [2, 9, 12, 21].

*We thank Matthias Troffaes for many valuable discussions and for suggesting the symmetric form of Coletti-Scozzafava's condition. Thanks also to Andrea Capotorti and Barbara Vantaggi for valuable comments on an earlier version. We acknowledge support from the *Fundação de Amparo à Pesquisa do Estado de São Paulo* (FAPESP) at various stages of this research.

Received by the editors: 11 October 2007.
Accepted for publication: 12 February 2008.

The goal of this paper is to compare concepts of independence for events and variables in the context of full conditional measures. Our strategy is to evaluate concepts of independence by the *graphoid* properties they satisfy (Section 3 reviews the theory of graphoids). This strategy is motivated by two observations. First, the graphoid properties have been often advocated as a compact set of properties that any concept of independence should satisfy. Even though some of the graphoid properties may have more limited scope than others, they offer a good starting point for discussions of independence. Second, the graphoid properties are useful in proving results about conditional probabilities, graphs, lattices, and other models [19]. In Sections 4 and 5 we analyze existing and new concepts of independence. We show that several key graphoid properties can fail due to null events.

2   Full Conditional Measures

In this paper, in order to avoid controversies about countable additivity for probability, we restrict ourselves to finite state spaces: $\Omega = \{\omega_1, \ldots, \omega_N\}$; any subset of Ω is an event. We use A, B, C to denote events and W, X, Y, Z to denote (sets of) random variables; by $A(X)$, $B(Y)$, $C(X)$ and $D(Y)$ we denote events defined either by X or by Y. Events such as $\{X = x\} \cap \{Y \neq y\} \cap \{Z = z\}$ are denoted simply as $\{X = x, Y \neq y, Z = z\}$.

A probability measure is a set function $P : 2^\Omega \to \mathbb{R}$ such that $P(\Omega) = 1$, $P(A) \geq 0$ for all A, and $P(A \cup B) = P(A) + P(B)$ for disjoint A and B. Given a probability measure, the probability of A conditional on B is usually defined to be $P(A \cap B)/P(B)$ when $P(B) > 0$; conditional probability is not defined if B is a null event. *Stochastic independence* of events A and B requires that $P(A \cap B) = P(A)P(B)$; or equivalently that $P(A|B) = P(A)$ when $P(B) > 0$. *Conditional stochastic independence* of events A and B given event C requires that $P(A \cap B|C) = P(A|C)P(B|C)$ if C is non-null. These concepts of independence can be extended to sets of events and to random variables by requiring more complex factorizations [22].

A different theory of probability ensues if we take conditional probability as a truly primitive concept, as already advocated by Keynes [35] and de Finetti [20]. The first question is the domain of probabilistic assessments. Rényi [47] investigates the general case where $P : \mathcal{B} \times \mathcal{C} \to \mathbb{R}$, where $\mathcal{B}$ is a Boolean algebra and $\mathcal{C}$ is an arbitrary subset of $\mathcal{B}$ (Rényi also requires σ-additivity). Popper considers a similar set-up [45]. Here we focus on $P : \mathcal{B} \times (\mathcal{B} \setminus \varnothing) \to \mathbb{R}$, where $\mathcal{B}$ is again a Boolean algebra [37], such that for every event $C \neq \varnothing$:

1. $P(C|C) = 1$;

2. $P(A|C) \geq 0$ for all A;

3. $P(A \cup B|C) = P(A|C) + P(B|C)$ for all disjoint A and B;

	A	A^{c}
B	$\lfloor\beta\rfloor_1$	α
B^{c}	$\lfloor 1-\beta\rfloor_1$	$1-\alpha$

TABLE 1. A simple full conditional measure $(\alpha, \beta \in (0,1))$ where $\mathrm{P}(A) = 0$ and $\mathrm{P}(B|A) = \beta$.

4. $\mathrm{P}(A\cap B|C) = \mathrm{P}(A|B\cap C)\mathrm{P}(B|C)$ for all A and B such that $B\cap C \neq \varnothing$.

This fourth axiom is often stated as $\mathrm{P}(A|C) = \mathrm{P}(A|B)\mathrm{P}(B|C)$ when $A \subseteq B \subseteq C$ and $B \neq \varnothing$ [4]. We refer to such a P as a *full conditional measure*, following Dubins [23]; there are other names in the literature, such as *conditional probability measure* [37] and *complete conditional probability system* [42]. Whenever the conditioning event C is equal to Ω, we suppress it and write the "unconditional" probability $\mathrm{P}(A)$ instead of $\mathrm{P}(A|\Omega)$.

Full conditional measures place no restrictions on conditioning on null events. If B is null, the constraint $\mathrm{P}(A\cap B) = \mathrm{P}(A|B)\mathrm{P}(B)$ is trivially true, and $\mathrm{P}(A|B)$ must be defined separately from $\mathrm{P}(B)$ and $\mathrm{P}(A\cap B)$. For any two events A and B, indicate by $A \gg B$ the fact that $\mathrm{P}(B|A \cup B) = 0$. Then we can partition Ω into events $L_0, \ldots, L_K$, where $K \leq N$, such that $L_i \gg L_{i+1}$ for $i \in \{0, \ldots, K-1\}$ if $K > 0$. Each event L_i is a "layer" of P, and the decomposition in layers always exists for any finite algebra. Coletti and Scozzafava denote by $\circ(A)$ the index i of the first layer L_i such that $\mathrm{P}(A|L_i) > 0$; they propose the convention $\circ(\varnothing) = \infty$ [12]. They also refer to $\circ(A)$ as the *zero-layer* of A; here we will use the term *layer level* of A for the same purpose. Note that some authors use a different terminology, where the ith "layer" is $\bigcup_{j=i}^K L_j$ rather than L_i [12, 37].

Coletti and Scozzafava also define the conditional layer number $\circ(A|B)$ as $\circ(A \cap B) - \circ(B)$ (defined only if $\circ(B)$ is finite).

Any full conditional measure can be represented as a sequence of strictly positive probability measures $\mathrm{P}_0, \ldots, \mathrm{P}_K$, where the support of P_i is restricted to L_i; that is, $\mathrm{P}_i : 2^{L_i} \to \mathbb{R}$. This result is proved assuming complete assessments in general spaces (not just finite) by Krauss [37] and Dubins [23], and it has been derived for partial assessments by Coletti [10, 12].

We have, for events A, B:

- $\mathrm{P}(B|A) = \mathrm{P}(B|A \cap L_{\circ(A)})$ [4, Lemma 2.1a].

- $\circ(A \cup B) = \min(\circ(A), \circ(B))$.

- Either $\circ(A) = 0$ or $\circ(A^{\mathrm{c}}) = 0$.

The following simple result will be useful later.

Lemma 2.1. Consider two random variables X and Y, event $A(X)$ defined by X and event $B(Y)$ defined by Y such that $A(X) \cap B(Y) \neq \varnothing$. If $\mathrm{P}(Y = y|\{X = x\} \cap B(Y)) = \mathrm{P}(Y = y|B(Y))$ for every $x \in A(X)$ such that $\{X = x\} \cap B(Y) \neq \varnothing$, then $\mathrm{P}(Y = y|A(X) \cap B(Y)) = \mathrm{P}(Y = y|B(Y))$.

Proof. We have (all summations run over $\{x \in A(X) : \{X = x\} \cap B(Y) \neq \varnothing\}$):

$$\mathrm{P}(Y = y|A(X) \cap B(Y)) =$$
$$= \sum \mathrm{P}(X = x, Y = y|A(X) \cap B(Y))$$
$$= \sum \mathrm{P}(Y = y|\{X = x\} \cap A(X) \cap B(Y)) \times \mathrm{P}(X = x|A(X) \cap B(Y))$$
$$= \sum \mathrm{P}(Y = y|\{X = x\} \cap B(Y))\mathrm{P}(X = x|A(X) \cap B(Y))$$
$$= \sum \mathrm{P}(Y = y|B(Y))\mathrm{P}(X = x|A(X) \cap B(Y))$$
$$= \mathrm{P}(Y = y|B(Y)) \sum \mathrm{P}(X = x|A(X) \cap B(Y))$$
$$= \mathrm{P}(Y = y|B(Y)).$$

Q.E.D.

The following notation will be useful. If A is such that $\circ(A) = i$ and $\mathrm{P}(A|L_i) = p$, we write $\lfloor p \rfloor_i$. If $\circ(A) = 0$ and $\mathrm{P}(A) = p$, we simply write p instead of $\lfloor p \rfloor_0$. Table 1 illustrates this notation.

There are several decision theoretic derivations of full conditional measures. The original arguments of de Finetti concerning called-off gambles [20] have been formalized in several ways [34, 46, 57, 58]. Derivations based on axioms on preferences have also been presented, both by Myerson [42] and by Blume *et al.* [7]. The last derivation is particularly interesting as it is based on non-Archimedean preferences and lexicographic preferences.

3   Graphoids

If we read $(X \perp\!\!\!\perp Y | Z)$ as "variable X is stochastically independent of variable Y given variable Z," then the following properties are true:

Symmetry: $(X \perp\!\!\!\perp Y | Z) \Rightarrow (Y \perp\!\!\!\perp X | Z)$

Decomposition: $(X \perp\!\!\!\perp (W, Y) | Z) \Rightarrow (X \perp\!\!\!\perp Y | Z)$

Weak union: $(X \perp\!\!\!\perp (W, Y) | Z) \Rightarrow (X \perp\!\!\!\perp W | (Y, Z))$

Contraction: $(X \perp\!\!\!\perp Y | Z)$ & $(X \perp\!\!\!\perp W | (Y, Z)) \Rightarrow (X \perp\!\!\!\perp (W, Y) | Z)$

Instead of interpreting ⊥⊥ as stochastic independence, we could take this relation to indicate an abstract concept of independence. The properties just outlined are then referred to as the *graphoid properties*, and any three-place relation that satisfies these properties is called a *graphoid*. Note that we are following terminology proposed by Geiger *et al.* in [27]; the term "graphoid" often means slightly different concepts [18, 43]. In fact, often the four properties just listed are called *semi-graphoid* properties, and the term "graphoid" is reserved to a relation that satisfies the semi-graphoid properties plus:

Intersection: $(X \perp\!\!\!\perp W \,|\, (Y, Z))$ & $(X \perp\!\!\!\perp Y \,|\, (W, Z)) \Rightarrow (X \perp\!\!\!\perp (W, Y) \,|\, Z)$

As the intersection property can already fail for stochastic independence in the presence of null events [14], it is less important than the other properties in the context of the present paper. But the intersection property is important for understanding Basu's First Theorem of statistical inference, as shown by San Martin *et al.* [39].

Finally, the following property is sometimes presented together with the previous ones [43]:

Redundancy: $(X \perp\!\!\!\perp Y \,|\, X)$

If we interpret W, X, Y and Z as *sets* of variables, redundancy implies that any property that is valid for disjoint sets of variables is also valid in general — because given symmetry, redundancy, decomposition and contraction, $(X \perp\!\!\!\perp Y \,|\, Z) \Leftrightarrow (X \backslash Z \perp\!\!\!\perp Y \backslash Z \,|\, Z)$ as noted by Pearl [43].

Graphoids offer a compact and intuitive abstraction of independence. As an example of application, several key results in the theory of Bayesian networks can be proved just using the graphoid properties, and are consequently valid for many possible generalizations of Bayesian networks [27]. Several authors have employed the graphoid properties as a benchmark to evaluate concepts of independence [3, 52, 55]; we follow the same strategy in this paper.

4   Epistemic and coherent irrelevance and independence

Because conditional probabilities are defined even on null events, we might consider a concise definition of independence: events A and B are independent iff $P(A|B) = P(A)$. However, this definition is not entirely satisfactory because it guarantees neither $P(A|B^c) = P(A)$ nor $P(B|A) = P(B)$ (failure of symmetry can be observed in Table 1). In this section we study the graphoid properties of three concepts of irrelevance/independence that attempt to correct these deficiencies. We collect results on *epistemic* and *strong coherent* irrelevance/independence, and then we explore the new concepts of *weak coherent* irrelevance/independence.

4.1   Epistemic irrelevance/independence

Keynes faced the problem of non-symmetric independence, in his theory of probability, by defining first a concept of *irrelevance* and then "symmetrizing" it [35]: B is irrelevant to A iff $\mathrm{P}(A|B) = \mathrm{P}(A)$; A and B are independent iff A is irrelevant to B and B is irrelevant to A. Walley strenghtened Keynes' definitions in his theory of imprecise probabilities: B is irrelevant to A iff $\mathrm{P}(A|B) = \mathrm{P}(A|B^c) = \mathrm{P}(A)$; independence is the symmetrized concept [56].[1] Note that Crisma has further strenghtened Walley's definitions by requiring logical independence [16] (we later discuss logical independence in more detail).

We follow Walley in using *epistemic irrelevance* of B to A to mean

$$\mathrm{P}(A|B) = \mathrm{P}(A) \text{ if } B \neq \varnothing \qquad \text{and} \qquad \mathrm{P}(A|B^c) = \mathrm{P}(A) \text{ if } B^c \neq \varnothing. \quad (4.1)$$

Epistemic independence refers to the symmetrized concept. Clearly epistemic irrelevance/independence can be extended to sets of events, random variables, and to concepts of conditional irrelevance/independence [56]. We wish to focus on:

Definition 4.1. Random variables X are epistemically irrelevant to random variables Y conditional on random variables Z, denoted by

$$(X \text{ EIR } Y \mid Z),$$

if $\mathrm{P}(Y = y|\{X = x, Z = z\}) = \mathrm{P}(Y = y|Z = z)$ for all values x, y, z whenever these probabilities are defined.

Epistemice independence, denoted using similar triplets with the symbol EIN , is the symmetrized concept.

We now consider the relationship between these concepts and the graphoid properties. Because irrelevance is not symmetric, there are several possible versions of the properties that might be of interest. For example, two different versions of weak union are $(X \text{ EIR } (W, Y) \mid Z) \Rightarrow (X \text{ EIR } W \mid (Y, Z))$ and $((W, Y) \text{ EIR } X \mid Z) \Rightarrow (W \text{ EIR } X \mid (Y, Z))$, and there are two additional possible versions. Decomposition also has four versions, while contraction and intersection have eight versions each. We single out two versions for each property, which we call the *direct* and the *reverse* versions. The direct version is obtained by writing the property as initially stated in Section 3, just replacing ⊥⊥ by EIR . The reverse version is obtained by switching every statement of irrelevance. Thus we have given respectively the

---

[1]Levi has also proposed $\mathrm{P}(A|B) = \mathrm{P}(A)$ as a definition of *irrelevance*, without considering the symmetrized concept [38]. Both Levi's and Walley's definitions are geared towards sets of full conditional measures, but clearly they specialize to a single full conditional measure.

direct and reverse versions of weak union in this paragraph (similar distinctions have appeared in the literature for various concepts of irrelevance [15, 24, 40, 52, 55]).

The following proposition relates epistemic irrelevance/independence with the graphoid properties (several results in the proposition can be extracted from Vantaggi's results [52]).

Proposition 4.2. Epistemic irrelevance satisfies the graphoid properties of direct and reverse redundancy, direct and reverse decomposition, reverse weak union, and direct and reverse contraction. If W and Y are logically independent, then epistemic irrelevance satisfies reverse intersection. All other versions of the graphoid properties and intersection fail for epistemic irrelevance. Epistemic independence satisfies symmetry, redundancy, decomposition and contraction — weak union and intersection fail for epistemic independence.

Proof. For epistemic irrelevance, the proof of direct and reverse redundancy, direct and reverse decomposition, reverse weak union and reverse contraction is obtained from the proof of Theorem 5.4, by taking $A(X) = B(Y) = \Omega$. For direct contraction $((X \text{ EIR } Y \mid Z) \ \& \ (X \text{ EIR } W \mid (Y, Z)) \Rightarrow (X \text{ EIR } (W, Y) \mid Z))$, consider that if $(X, Z) \neq \varnothing$,

$$
\begin{aligned}
\mathrm{P}(W, Y | X, Z) &= \mathrm{P}(W | X, Y, Z)\mathrm{P}(Y | X, Z) \\
&= \mathrm{P}(W | Y, Z)\mathrm{P}(Y | Z) \\
&= \mathrm{P}(W, Y | Z),
\end{aligned}
$$

where the term $\mathrm{P}(W | X, Y, Z)\mathrm{P}(Y | X, Z)$ is only defined when (X, Y, Z) is nonempty (if it is empty, then the other equalities are valid because both sides are equal to zero). All other versions of graphoid properties fail for epistemic irrelevance, as shown by measures in Tables 2, 3, 7, and 8. For epistemic independence, symmetry is true by definition; redundancy, decomposition and contraction follow from their direct and reverse versions for epistemic irrelevance; Table 2 displays failure of weak union, and Table 3 displays failure of intersection.   Q.E.D.

4.2 Strong coherent irrelevance/independence

Coletti and Scozzafava have proposed a concept of independence that explicitly deals with layers [11, 13, 52]. They define the following condition:

Definition 4.3. The *Coletti-Scozzafava condition* on (B, A) holds iff $B \neq \varnothing \neq B^c$ and $\circ(A | B) = \circ(A | B^c)$ and $\circ(A^c | B) = \circ(A^c | B^c)$, where these layer numbers are computed with respect to probabilities over the events $\{A \cap B, A \cap B^c, A^c \cap B, A^c \cap B^c\}$.

	$w_0 y_0$	$w_1 y_0$	$w_0 y_1$	$w_1 y_1$
x_0	α	$\lfloor \beta \rfloor_2$	$1 - \alpha$	$\lfloor 1 - \beta \rfloor_2$
x_1	$\lfloor \alpha \rfloor_1$	$\lfloor \gamma \rfloor_3$	$\lfloor 1 - \alpha \rfloor_1$	$\lfloor 1 - \gamma \rfloor_3$

TABLE 2. Failure of direct weak union for epistemic irrelevance/independence $(\alpha, \beta, \gamma \in (0,1)$, with $\alpha \neq \beta \neq \gamma)$. The full conditional measure in the table satisfies $(X \text{ EIN } (W, Y))$ but fails $(X \text{ EIR } Y \mid W)$.

	$w_0 y_0$	$w_1 y_0$	$w_0 y_1$	$w_1 y_1$
x_0	$\lfloor 1 \rfloor_3$	$\lfloor \beta \rfloor_1$	$\lfloor (1 - \beta) \rfloor_1$	$\lfloor 1 \rfloor_5$
x_1	$\lfloor 1 \rfloor_2$	α	$(1 - \alpha)$	$\lfloor 1 \rfloor_4$

TABLE 3. Failure of versions of intersection for epistemic irrelevance/independence $(\alpha, \beta \in (0,1)$, with $\alpha \neq \beta)$. The full conditional measure in the table satisfies $(X \text{ EIN } W \mid Y)$ and $(X \text{ EIN } Y \mid W)$, but not $(X \text{ EIR } (W, Y))$.

	A	A^c		A	A^c
B	$\lfloor 1 \rfloor_1$	α	B	1	$\lfloor 1 \rfloor_2$
B^c	$\lfloor 1 \rfloor_2$	$1 - \alpha$	B^c	$\lfloor 1 \rfloor_1$	$\lfloor 1 \rfloor_4$

TABLE 4. The Coletti-Scozzafava condition, where $\alpha \in (0,1)$.

The Coletti-Scozzafava condition focuses only on the four events $\{A \cap B, A \cap B^c, A^c \cap B, A^c \cap B^c\}$; consequently, it never deals with layer numbers larger than 3.[2] A situation that violates the Coletti-Scozzafava condition is depicted in Table 4 (left). The example in Table 4 (right) fails the following result by Coletti and Scozzava [12, Theorem 17]: if A and B are logically independent, then $P(A|B) = P(A|B^c)$ and $P(B|A) = P(B|A^c)$ imply $\circ(A^c|B) = \circ(A^c|B^c)$. But this result is true when we focus on the restricted set of events $\{A \cap B, A \cap B^c, A^c \cap B, A^c \cap B^c\}$.

Coletti and Scozzafava then define independence of B to A as: satisfaction of the Coletti-Scozzafava condition on (B, A) *plus* epistemic irrelevance of B to A. The concept is not symmetric (Table 1).

Another important aspect of the Coletti-Scozzafava condition is that if B (or B^c) is empty, then B is not deemed irrelevant to any other event. Coletti and Scozzafava argue that their condition deals adequately with logical independence, as follows. Consider a table containing layer numbers for events $\{A \cap B, A \cap B^c, A^c \cap B, A^c \cap B^c\}$. Every entry can be either

---

[2]We thank Andrea Capotorti for bringing this fact to our attention.

finite or infinite; thus we have 16 configurations insofar as this distinction is concerned. One of these configurations is impossible (the one with four infinite entries). The other configurations can be divided in the cases depicted in Table 5. The Coletti-Scozzafava condition blocks the irrelevance of B to A in the second, third, fourth, sixth and seventh tables. Most cases of logical independence are thus removed. (However, the fifth table does not fail the Coletti-Scozzafava condition even though displays a situation of logical dependence,[3] as noted by Coletti and Scozzafava [12, Proposition 2].)

Vantaggi extends the main ideas in Coletti and Scozzafava's concept of independence by proposing a condition that can be stated as follows:

Definition 4.4. The *conditional Coletti-Scozzafava condition* on variables (Y, X) given variables Z holds iff for all values x, y, z,

$$\{Y = y, Z = z\} \neq \varnothing \quad \text{and} \quad \{Y \neq y, Z = z\} \neq \varnothing,$$

and

$$\circ(X = x | Y = y, Z = z) = \circ(X = x | Y \neq y, Z = z)$$

and

$$\circ(X \neq x | Y = y, Z = z) = \circ(X \neq x | Y \neq y, Z = z),$$

where these layer numbers are computed with respect to $\{\{X = x, Y = y, Z = z\}, \{X = x, Y \neq y, Z = z\}, \{X \neq x, Y = y, Z = z\}, \{X \neq x, Y \neq y, Z = z\}\}$.

Vantaggi then proposes a concept, here referred to as *strong coherent irrelevance*:[4] Y is strongly coherently irrelevant to X given Z if (i) Y is epistemically irrelevant to X given Z, and (ii) the conditional Coletti-Scozzafava condition holds on (Y, X) given Z [52, Definition 7.1]. This is a very stringent concept, as strong coherent irrelevance requires Y and Z to be logically independent.

As shown by Vantaggi [52], strong coherent irrelevance fails symmetry, direct and reverse redundancy, and direct weak union. Her results imply that a symmetrized concept of strong coherent independence fails redundancy and weak union (Table 2). Note also that strong coherent irrelevance fails direct intersection, as shown by Table 3.

---

[3]We thank Barbara Vantaggi for bringing this fact to our attention.

[4]It should be noted that Vantaggi's concepts of independence correspond to irrelevance in our terminology (she also studies special cases where symmetry holds [54]). Also, Vantaggi's writing of properties is different from ours; for example, her reverse weak union property is our direct weak union property.

	A	A^c
B	a	b
B^c	c	d

	A	A^c
B	a	∞
B^c	c	d

	A	A^c
B	a	∞
B^c	∞	d

	A	A^c
B	∞	b
B^c	c	∞

	A	A^c
B	a	∞
B^c	c	∞

	A	A^c
B	a	b
B^c	∞	∞

	A	A^c
B	a	∞
B^c	∞	∞

TABLE 5. Cases for the Coletti-Scozzafava and layer conditions.

4.3   The layer condition and weak coherent irrelevance/independence

In this sub-section we consider a condition on layer numbers that relaxes the Coletti-Scozzafava condition in two ways. First, we remove the restriction on the set of events that must be taken into account to determine layer numbers. Second, we replace conjunction with material implication in the case B (or B^c) is empty, as this replacement will lead to a host of interesting properties. In Section 6 we comment on our reasons to make these changes and on their consequences.

Definition 4.5. The *layer condition* on (B, A) holds iff whenever $B \neq \varnothing \neq B^c$ then $\circ(A|B) = \circ(A|B^c)$ and $\circ(A^c|B) = \circ(A^c|B^c)$.

The first thing to note is that the layer condition is symmetric.[5]

Proposition 4.6. If A and B satisfy the layer condition on (B, A), they satisfy the layer condition on (A, B).

Proof. Define $a = \circ(A \cap B)$; $b = \circ(A \cap B^c)$; $c = \circ(A^c \cap B)$; $d = \circ(A^c \cap B^c)$. Each one of these four layer levels may be finite or infinite. There are thus 16 situations; one is impossible and six others always violate the layer condition. Suppose then that no layer level is infinite and assume $a = 0$, as one of the four entries must be 0. The first table in Table 5 illustrates this case. Regardless of the value of c, $\circ(B^c) = c$ because $\circ(A \cap B^c) - \circ(B^c) = 0$ by hypothesis. Then $b - 0 = d - c$, thus $d = c + b$ as desired. Now, if $\circ(A \cap B) \neq 0$, then we can always re-label rows and columns so that the top left entry is zero, and the same reasoning follows. The remaining cases are illustrated by the fifth, sixth, and seventh tables in Table 5, where the result is immediate.   Q.E.D.

---

[5]The Coletti-Scozzavafa condition is not symmetric, as shown by the sixth table in Table 5.

Consequently, it is enough to indicate that two events satisfy the layer condition, without mentioning a "direction" (B, A) or (A, B). We have:

Proposition 4.7. Events A and B satisfy the layer condition iff $\circ(A|B) = \circ(A)$, $\circ(A|B^c) = \circ(A)$, $\circ(A^c|B) = \circ(A^c)$ and $\circ(A^c|B^c) = \circ(A^c)$ whenever the relevant quantities are defined.

Proof. Direct by verifying all tables in Table 5 (it may be necessary to re-label rows and columns to deal with the nine relevant situations discussed in the proof of Proposition 4.6).   Q.E.D.

The previous result directly implies equivalence of the layer condition and a more obviously symmetric condition:[6]

Corollary 4.8. Events A and B satisfy the layer condition iff

$$\circ(A \cap B) = \circ(A) + \circ(B), \qquad \circ(A \cap B^c) = \circ(A) + \circ(B^c),$$

$$\circ(A^c \cap B) = \circ(A^c) + \circ(B), \qquad \circ(A^c \cap B^c) = \circ(A^c) + \circ(B^c).$$

We now consider a version of the layer condition for random variables, clearly similar to Vantaggi's conditional Coletti-Scozzafava condition:[7]

Definition 4.9. The *conditional layer condition* on variables (Y, X) given variables Z holds iff for all values x, y, z, whenever $\{Y = y, Z = z\} \neq \varnothing$ and $\{Y \neq y, Z = z\} \neq \varnothing$, then

$$\circ(X = x|Y = y, Z = z) = \circ(X = x|Y \neq y, Z = z)$$

and

$$\circ(X \neq x|Y = y, Z = z) = \circ(X \neq x|Y \neq y, Z = z).$$

Before we propose concepts of irrelevance/independence based on the conditional layer condition, we examine some useful properties of this condition.

Proposition 4.10. The conditional layer condition is equivalent to

$$\circ(X = x|Y = y, Z = z) = \circ(X = x|Z = z) \qquad (4.2)$$

for all x, y, z such that $\{Y = y, Z = z\} \neq \varnothing$.

---

[6]This result was suggested to us by Matthias Troffaes.

[7]We note that Vantaggi's Definition 7.3 from [52] uses a condition that is very close to the conditional layer condition in Definition 4.9; the difference basically is that Vantaggi requires every layer number to be computed with respect to $\{\{X = x, Y = y, Z = z\}, \{X = x, Y \neq y, Z = z\}, \{X \neq x, Y = y, Z = z\}, \{X \neq x, Y \neq y, Z = z\}\}$.

Proof. Assume the conditional layer condition; using Proposition 4.7 for every $\{Z = z\}$, we obtain $\circ(X = x | Y = y, Z = z) = \circ(X = x | Z = z)$ for all x, y and z such that $\{Y = y, Z = z\} \neq \varnothing$. Now assume Expression (4.2), and denote by $B(Y)$ an event defined by Y such that $B(Y) \cap \{Z = z\} \neq \varnothing$. Then we have

$$
\begin{aligned}
\circ(X = x | B(Y) \cap \{Z = z\}) = \\
= \min_{y \in B(Y)} \circ(X = x, Y = y | Z = z) - \min_{y \in B(Y)} \circ(Y = y | Z = z) \\
= \min_{y \in B(Y)} \left(\circ(X = x | Y = y, Z = z) + \circ(Y = y | Z = z) \right) \\
- \min_{y \in B(Y)} \circ(Y = y | Z = z) \\
= \min_{y \in B(Y)} \left(\circ(X = x | Z = z) + \circ(Y = y | Z = z) \right) \\
- \min_{y \in B(Y)} \circ(Y = y | Z = z) \\
= \circ(X = x | Z = z) + \min_{y \in B(Y)} \circ(Y = y | Z = z) \\
- \min_{y \in B(Y)} \circ(Y = y | Z = z) \\
= \circ(X = x | Z = z) ,
\end{aligned}
$$

where the minima are taken with respect to values of Y that are logically independent of $\{Z = z\}$. Thus the first part of the conditional layer condition is satisfied. For the second part, note that

$$
\begin{aligned}
\circ(X \neq x | B(Y) \cap \{Z = z\}) &= \min_{x' \neq x} \circ(X = x' | B(Y) \cap \{Z = z\}) \\
&= \min_{x' \neq x} \circ(X = x' | Z = z) \\
&= \circ(X \neq x | Z = z) .
\end{aligned}
$$

Q.E.D.

A more obviously symmetric version of the conditional layer condition is:

Corollary 4.11. The conditional layer condition is equivalent to

$$\circ(X = x, Y = y | Z = z) = \circ(X = x | Z = z) + \circ(Y = y | Z = z) \qquad (4.3)$$

for all x, y, z such that $\{Z = z\} \neq \varnothing$.

Proof. The fact that Expression (4.3) implies Expression (4.2) is immediate from the definition of $\circ(X = x, Y = y | Z = z)$. To prove the converse, consider the two possible cases. For all y, z such that $\{Y = y, Z = z\} \neq \varnothing$,

Proposition 4.10 directly yields the result as we have $\circ(X = x, Y = y|Z = z) - \circ(Y = y|Z = z) = \circ(X = x|Z = z)$. If instead $\{Y = y, Z = z\} = \varnothing$, then $\circ(X = x, Y = y|Z = z) = \circ(Y = y|Z = z) = \infty$ and Expression (4.3) is satisfied regardless of $\circ(X = x|Z = z)$.   Q.E.D.

Denote by $(X \, \text{LC} \, Y \mid Z)$ the fact that X and Y satisfy the conditional layer condition given Z. It is interesting to note that this relation is a graphoid.

Proposition 4.12. The relation $(X \, \text{LC} \, Y \mid Z)$ satisfies symmetry, redundancy, decomposition and contraction; if X and Y are logically independent given Z, then intersection is satisfied.

Proof. **Symmetry** follows from Expression (4.3). In order to show **Redundancy** (i.e., $(Y \perp\!\!\!\perp X \mid X)$), we need to show that

$$\circ(X = x_1|Y = y, X = x_2) = \circ(X = x_1|X = x_2).$$

If $x_1 \neq x_2$, $\circ(X = x_1|Y = y, X = x_2) = \circ(X = x_1|X = x_2) = \infty$; if $x_1 = x_2$, we obtain

$$\circ(X = x_1, Y = y, X = x_1) - \circ(X = x_1, Y = y) = 0 =$$
$$= \circ(X = x_1, X = x_2) - \circ(X = x_1).$$

Let us now show **Decomposition** (i.e., $(X \perp\!\!\!\perp (W, Y) \mid Z) \Rightarrow (X \perp\!\!\!\perp W \mid Z)$). We have that

$$\circ(W = w, X = x|Z = z) =$$
$$= \min_y \circ(W = w, X = x, Y = y|Z = z)$$
$$= \min_y \circ(X = x|Z = z) + \circ(W = w, Y = y|Z = z)$$
$$= \circ(X = x|Z = z) + \min_y \circ(W = w, Y = y|Z = z)$$
$$= \circ(X = x|Z = z) + \circ(W = w|Z = z).$$

Moving on to **Weak union** (i.e., $((W, Y) \perp\!\!\!\perp X \mid Z) \Rightarrow (W \perp\!\!\!\perp X \mid (Y, Z)))$, we have by hypothesis that

$$\circ(X = x|W = w, Y = y, Z = z) = \circ(X = x|Z = z),$$

and from this (by **Decomposition**) we obtain $\circ(X = x|Y = y, Z = z) = \circ(X = x|Z = z)$; consequently

$$\circ(X = x|W = w, Y = y, Z = z) = \circ(X = x|Y = y, Z = z).$$

	$w_0 y_0$	$w_1 y_0$	$w_0 y_1$	$w_1 y_1$
x_0	0	∞	∞	0
x_1	1	∞	∞	2

TABLE 6. Failure of the intersection property (for the layer conditions) in the absence of logical independence. Entries in the table are layer levels; the four central entries denote empty events.

Concerning **Contraction**, we have

$$\circ(X = x | W = w, Y = y, Z = z) \;=\; \circ(X = x | Y = y, Z = z)$$
$$=\; \circ(X = x | Z = z).$$

Finally, for **Intersection** (i.e., $(W \perp\!\!\!\perp X \,|\, (Y, Z))$ & $(Y \perp\!\!\!\perp X \,|\, (W, Z)) \Rightarrow ((W, Y) \perp\!\!\!\perp X \,|\, Z))$ we use the fact that, due to the hypothesis of logical independence, we have

$$\circ(X = x | W = w, Z = z) \;=\; \circ(X = x | W = w, Y = y, Z = z)$$
$$=\; \circ(X = x | Y = y, Z = z)$$

for all (w, y). Then

$$\circ(X = x | Z = z) \;=\; \min_w \circ(X = x, W = w | Z = z)$$
$$=\; \min_w \left(\circ(X = x | W = w, Z = z) + \circ(W = w | Z = z) \right)$$
$$=\; \min_w \left(\circ(X = x | Y = y, Z = z) + \circ(W = w | Z = z) \right)$$
$$=\; \circ(X = x | Y = y, Z = z) + \min_w \circ(W = w | Z = z)$$
$$=\; \circ(X = x | Y = y, Z = z)$$
$$=\; \circ(X = x | W = w, Y = y, Z = z)$$

(because $\min_w \circ(W = w | Z = z) = 0$ and then using the hypothesis). The hypothesis of logical independence is necessary, as shown by Table 6.   Q.E.D.

We can now define concepts of irrelevance and independence that employ the conditional layer condition:

Definition 4.13. Random variables Y are *weakly coherently irrelevant* to random variables X given random variables Z, denoted by $(Y$ WCIR $X \mid Z)$, iff $(Y$ EIR $X \mid Z)$ and $(Y$ LC $X \mid Z)$. Random variables X and Y are *weakly coherently independent* given random variables Z, denoted by $(Y$ WCIN $X \mid Z)$, iff $(Y$ EIN $X \mid Z)$ and $(Y$ LC $X \mid Z)$.

Given our previous results, we easily obtain the following result.

Proposition 4.14. Weak coherent irrelevance satisfies the graphoid properties of direct and reverse redundancy, direct and reverse decomposition, reverse weak union, and direct and reverse contraction. If W and Y are logically independent, then weak coherent irrelevance satisfies reverse intersection. All other versions of the graphoid properties and intersection fail for weak coherent irrelevance. Weak coherent independence satisfies the graphoid properties of symmetry, redundancy, decomposition and contraction — weak union and intersection fail for weak coherent independence.

Proof. Equalities among probability values have been proved for Proposition 4.2, and equalities among layer levels have been proved for Proposition 4.12. All failures of symmetry, decomposition, weak union, contraction and intersection discussed in Proposition 4.2 are still valid, with the same examples.   Q.E.D.

We have thus examined three concepts of independence that fail the weak union and the intersection properties. The failure of intersection is not too surprising, as this property requires strictly positive probabilities even with the usual concept of stochastic independence. However, the failure of weak union leads to serious practical consequences. To give an example, consider the theory of Bayesian networks [43], where weak union is necessary to guarantee that a structured set of independence relations has a representation based on a graph, and to guarantee that independences can be read off a graph using the d-separation criterion. We intend to explore the relationships between Bayesian networks and full conditional measures in a future publication.

In the next section we examine concepts of irrelevance/independence that satisfy the weak union property.

5   Full irrelevance and independence

Even a superficial analysis of Table 2 suggests that epistemic and coherent independence fail to detect obvious dependences among variables: there is a clear disparity between the two rows, as revealed by conditioning on $\{W = w_1\}$. The problem is that epistemic independence is regulated by the "first active" layer, and it ignores the content of lower layers. Hammond has proposed a concept of independence that avoids this problem by requiring [31]:

$$P(A(X) \cap B(Y)|C(X) \cap D(Y)) = P(A(X)|C(X))P(B(Y)|D(Y)), \quad (5.1)$$

for all events $A(X)$, $C(X)$ defined by X, and all events $B(Y)$, $D(Y)$ defined by Y, such that $C(X) \cap D(Y) \neq \varnothing$. Hammond shows that this symmetric definition can be decomposed into two non-symmetric parts as follows.

Definition 5.1. Random variables X are h-irrelevant to random variables Y (denoted by $(X \text{ HIR } Y)$) iff $P(B(Y)|A(X) \cap D(Y)) = P(B(Y)|D(Y))$, for all events $B(Y)$, $D(Y)$ defined by Y, and all events $A(X)$ defined by X, such that $A(X) \cap D(Y) \neq \emptyset$.

If $(X \text{ HIR } Y)$ and $(Y \text{ HIR } X)$, then X and Y are *h-independent*. Expression (5.1) is equivalent to h-independence of X and Y (for one direction, take first $A(X) = C(X)$ and then $B(Y) = D(Y)$; for the other direction, note that $P(A(X) \cap B(Y)|C(X) \cap D(Y))$ is equal to the product $P(A(X)|B(Y) \cap C(X) \cap D(Y))P(B(Y)|C(X) \cap D(Y)))$.

We can extend Hammond's definition to conditional independence (a move that has not been made by Hammond himself; Halpern mentions a conditional version of Hammond's definition under the name of *approximate independence* [30]):

Definition 5.2. Random variables X are h-irrelevant to random variables Y conditional on random variables Z (denoted by $(X \text{ HIR } Y \mid Z)$) iff

$$P(B(Y)|\{Z = z\} \cap A(X) \cap D(Y)) = P(B(Y)|\{Z = z\} \cap D(Y)),$$

for all values z, all events $B(Y)$, $D(Y)$ defined by Y, and all events $A(X)$ defined by X, such that $\{Z = z\} \cap A(X) \cap D(Y) \neq \emptyset$.

The "symmetrized" concept is:

Definition 5.3. Random variables X and Y are h-independent conditional on random variables Z (denoted by $(X \text{ HIN } Y \mid Z)$) iff $(X \text{ HIR } Y \mid Z)$ and $(Y \text{ HIR } X \mid Z)$.

This symmetric concept of conditional h-independence is equivalent to (analogously to Expression (5.1)):

$$P(A(X) \cap B(Y)|\{Z = z\} \cap C(X) \cap D(Y)) = \qquad (5.2)$$
$$P(A(X)|\{Z = z\} \cap C(X))P(B(Y)|\{Z = z\} \cap D(Y)),$$

whenever $\{Z = z\} \cap C(X) \cap D(Y) \neq \emptyset$.

The definition of h-irrelevance can be substantially simplified: for random variables X, Y, and Z, $(X \text{ HIR } Y \mid Z)$ iff

$$P(Y = y|\{X = x, Z = z\} \cap D(Y)) = P(Y = y|\{Z = z\} \cap D(Y))$$

for all x, y, z and all events $D(Y)$ defined by Y such that $\{X = x, Z = z\} \cap D(Y) \neq \emptyset$ (directly from Lemma 2.1).

The positive feature of h-irrelevance is that it satisfies direct weak union, and in fact h-independence satisfies weak union. Unfortunately, both concepts face difficulties with contraction.

Theorem 5.4. H-irrelevance satisfies the graphoid properties of direct and reverse redundancy, direct and reverse decomposition, direct and reverse weak union, and reverse contraction. If W and Y are logically independent, then h-irrelevance satisfies reverse intersection. All other versions of the graphoid properties and intersection fail for h-irrelevance. H-independence satisfies the graphoid properties of symmetry, redundancy, decomposition and weak union — contraction and intersection fail for h-independence.

Proof. Denote by $A(X)$, $B(Y)$ arbitrary events defined by X and Y respectively, chosen such that if they appear in conditioning, they are not logically incompatible with other events. We abbreviate the set $\{W = w\}$ by w, and likewise use x for $\{X = x\}$, y for $\{Y = y\}$, z for $\{Z = z\}$.

Symmetry fails for h-irrelevance as shown by Table 1. **Direct redundancy** (i.e., $(X \text{ HIR } Y \mid X)$) holds because

$$P(Y = y | \{X = x_1, X = x_2\} \cap B(Y)) = P(Y = y | \{X = x_1\} \cap B(Y)),$$

when $x_1 = x_2$ (and $\{X = x_1, X = x_2\} = \varnothing$ otherwise). Furthermore, **Reverse redundancy** (i.e., $(Y \text{ HIR } X \mid X)$) holds because

$$P(X = x_1 | \{Y = y, X = x_2\} \cap A(X)) = P(X = x_1 | \{X = x_2\} \cap A(X)) = 0$$

if $x_1 \neq x_2$ and

$$P(X = x_1 | \{Y = y, X = x_2\} \cap A(X)) = P(X = x_1 | \{X = x_2\} \cap A(X)) = 1$$

if $x_1 = x_2$. We also see that **Direct decomposition** holds as

$$
\begin{aligned}
P(y | \{x, z\} \cap B(Y)) &= \sum_w P(w, y | \{x, z\} \cap B(Y)) \\
&= \sum_w P(w, y | \{z\} \cap B(Y)) \\
&= P(y | \{z\} \cap B(Y)).
\end{aligned}
$$

In the argument for **Reverse decomposition** (i.e., $((W, Y) \text{ HIR } X \mid Z) \Rightarrow (Y \text{ HIR } X \mid Z)$) note that summations over values of W need only include values such that $\{w, y, z\} \cap A(X) \neq \varnothing$. Then we see that **Reverse decomposition** holds because

$$
\begin{aligned}
P(x | \{y, z\} \cap A(X)) &= \sum_w P(w, x | \{y, z\} \cap A(X)) \\
&= \sum_w P(x | \{w, y, z\} \cap A(X)) P(w | \{y, z\} \cap A(X)) \\
&= \sum_w P(x | \{z\} \cap A(X)) P(w | \{y, z\} \cap A(X))
\end{aligned}
$$

$$= P(x|\{z\} \cap A(X)) \sum_w P(w|\{y, z\} \cap A(X))$$

$$= P(x|\{z\} \cap A(X)).$$

Since $\{w\} \cap B(Y)$ is an event defined by (W, Y), we consequently have that

$$
\begin{aligned}
P(y|\{w, x, z\} \cap B(Y)) &= P(w, y|\{x, z\} \cap (\{w\} \cap B(Y))) \\
&= P(w, y|\{z\} \cap (\{w\} \cap B(Y))) \\
&= P(y|\{w, z\} \cap B(Y)),
\end{aligned}
$$

and thus **Direct weak union** holds. Furthermore, **Reverse weak union** holds because

$$P(x|\{w, y, z\} \cap A(X)) = P(x|\{z\} \cap A(X))$$

is true by hypothesis and $P(x|\{z\} \cap A(X)) = P(x|\{y, z\} \cap A(X))$ by **Reverse decomposition**. To see that **Reverse contraction** holds, we just check that $P(x|\{w, y, z\} \cap A(X)) = P(x|\{y, z\} \cap A(X)) = P(x|\{z\} \cap A(X))$. In order to see that **Reverse intersection** holds, we check that, due to the hypothesis of logical independence,

$$P(x|\{w, z\} \cap A(X)) = P(x|\{w, y, z\} \cap A(X)) = P(x|\{y, z\} \cap A(X))$$

for all (w, y). Thus we can write

$$
\begin{aligned}
P(x|\{z\} \cap A(X)) &= \sum_w P(x, w|\{z\} \cap A(X)) \\
&= \sum_w P(x|\{w, z\} \cap A(X)) P(w|\{z\} \cap A(X)) \\
&= \sum_w P(x|\{y, z\} \cap A(X)) P(w|\{z\} \cap A(X)) \\
&= P(x|\{y, z\} \cap A(X)) \sum_w P(w|\{z\} \cap A(X)) \\
&= P(x|\{y, z\} \cap A(X)) \\
&= P(x|\{w, y, z\} \cap A(X)).
\end{aligned}
$$

All other versions of graphoid properties fail for h-irrelevance, as shown by measures in Tables 2, 3, 7, and 8.

Now consider the "symmetrized" concept of h-independence. Symmetry is true by definition; redundancy, decomposition and contraction come from their direct and reverse versions for h-irrelevance. Table 2 displays failure of contraction, and Table 3 displays failure of intersection for h-independence.

Q.E.D.

Note that Table 2 is now responsible for failure of direct contraction, as now $(X \text{ HIR } Y)$ and $(X \text{ HIR } W \mid Y)$ but not $(X \text{ HIR } (W, Y))$.

It is natural to consider the strenghtening of h-independence with the conditional layer condition. The first question is whether or not to strenghten the conditional layer condition itself. We might consider the following condition:

$$\circ(X = x \mid \{Y = y, Z = z\} \cap A(X)) = \circ(X = x \mid \{Z = z\} \cap A(X)) \quad (5.3)$$

for all x, y, z and every event $A(X)$ defined by X such that and $\{Y = y, Z = z\} \cap A(X) \neq \varnothing$. As shown by the next result, this condition is implied by the conditional layer condition.

Proposition 5.5. If Expression (4.2) holds for all x, y, z such that and $\{Y = y, Z = z\} \neq \varnothing$, then Expression (5.3) holds for all x, y, z and every event $A(X)$ defined by X such that $\{Y = y, Z = z\} \cap A(X) \neq \varnothing$.

Proof. If $x \notin A(X)$, then the relevant layer levels are both equal to infinity. Suppose then that $x \in A(X)$. Using the abbreviations adopted in the proof of Theorem 5.7 for events such as $\{X = x\}$, we have:

$$
\begin{aligned}
\circ(x \mid \{y, z\} \cap A(X)) &= \circ(\{x\} \cap A(X) \mid y, z) - \circ(A(X) \mid y, z) \\
&= \circ(\{x\} \cap A(X) \mid y, z) - \min_{x' \in A(X)} \circ(x' \mid y, z) \\
&= \circ(\{x\} \cap A(X) \mid z) - \min_{x' \in A(X)} \circ(x' \mid z) \\
&= \circ(x \mid \{z\} \cap A(X)) .
\end{aligned}
$$

Q.E.D.

We propose the following definition.

Definition 5.6. Random variables X are *fully* irrelevant to random variables Y conditional on random variables Z (denoted $(X \text{ FIR } Y \mid Z)$) iff

$$P(Y = y \mid \{X = x, Z = z\} \cap B(Y)) = P(Y = y \mid \{Z = z\} \cap B(Y)),$$

$$\circ(Y = y \mid X = x, Z = z) = \circ(Y = y \mid Z = z) ,$$

for all x, y, z, and all events $B(Y)$ defined by Y such that $\{X = x, Z = z\} \cap B(Y) \neq \varnothing$.

Full independence is the symmetrized concept. Theorem 5.7 and Proposition 4.14 then imply the following result.

Theorem 5.7. Full irrelevance satisfies the graphoid properties of direct and reverse redundancy, direct and reverse decomposition, direct and reverse weak union, and reverse contraction. If W and Y are logically independent, then full irrelevance satisfies reverse intersection. All other versions of the graphoid properties and intersection fail for full irrelevance. Full independence satisfies the graphoid properties of symmetry, redundancy, decomposition and weak union — contraction and intersection fail for full independence.

We have thus examined two concepts of independence that fail the contraction and intersection properties. While failure of intersection is not surprising, the failure of contraction has important consequences — for example, contraction is needed in the theory of Bayesian networks for essentially the same reasons that weak union is needed.

It must be noted that only one of direction of contraction fails. Specifically, direct contraction fails, and direct contraction is a much less compelling than most other graphoid properties. While reverse contraction convincingly stands for the fact that "if we judge W irrelevant to X after learning some irrelevant information Y, then W must have been irrelevant [to X] before we learned Y" [43]; a similar reading of direct contraction reveals a much less intuitive property. On this account, it seems that h-irrelevance is more appropriate than epistemic irrelevance, and that full irrelevance is more appropriate than weak coherent irrelevance — and likewise for the corresponding independence concepts.

6 Conclusion

In this paper we have examined properties of full conditional measures, focusing on concepts of irrelevance and independence. We started by reviewing the concepts of epistemic and strong coherent irrelevance/independence. We then introduced the layer condition and the related concept of weak coherent irrelevance/independence (Definitions 4.9 and 4.13), and derived their graphoid properties (Propositions 4.12 and 4.14). We then moved to concepts of irrelevance/independence that satisfy the weak union property. We have:

- presented an analysis of Hammond's concepts of irrelevance and independence with respect to graphoid properties (Theorem 5.4) — in fact, we note that Hammond and others have not attempted to study *conditional* irrelevance and independence (our Definitions 5.2 and 5.3);

- introduced the definition of full irrelevance and independence, and presented the analysis of their graphoid properties (Definition 5.6 and Theorem 5.7).

	$w_0 y_0$	$w_1 y_0$	$w_0 y_1$	$w_1 y_1$
x_0	$\lfloor \gamma\beta \rfloor_1$	$\lfloor \gamma(1-\beta) \rfloor_1$	$\alpha\beta$	$\alpha(1-\beta)$
x_1	$\lfloor (1-\gamma)\beta \rfloor_1$	$\lfloor (1-\gamma)(1-\beta) \rfloor_1$	$(1-\alpha)\beta$	$(1-\alpha)(1-\beta)$

	$w_0 y_0$	$w_1 y_0$	$w_0 y_1$	$w_1 y_1$
x_0	$\lfloor \gamma\beta \rfloor_1$	$\lfloor \gamma(1-\beta) \rfloor_1$	$\lfloor (1-\gamma)\beta \rfloor_1$	$\lfloor (1-\gamma)(1-\beta) \rfloor_1$
x_1	$\alpha\beta$	$\alpha(1-\beta)$	$(1-\alpha)\beta$	$(1-\alpha)(1-\beta)$

TABLE 7. Failure of versions of decomposition, weak union, contraction, and intersection for epistemic irrelevance ($\alpha, \beta, \gamma \in (0,1)$, with $\alpha \neq \beta \neq \gamma \neq \alpha$). The full conditional measure in the top table satisfies (X EIR (W, Y)) but it fails (Y EIR X) (version of decomposition) and it fails (Y EIR $X \mid W$) (version of weak union); it satisfies (X EIR Y) and (W EIN $X \mid Y$) but it fails (($W, Y)$ EIR X) (two versions of contraction); it also satisfies (X EIR $Y \mid W$) (two versions of intersection, and by switching W and Y, another version of intersection). The full conditional measure in the bottom table satisfies (($W, Y)$ EIR X) but it fails (X EIR Y) (version of decomposition) and it fails (X EIR $Y \mid W$) (version of weak union); it satisfies (Y EIR X) and (W EIN $X \mid Y$), but it fails (X EIR (W, Y)) (two versions of contraction).

The results in this paper show that there are subtle challenges in combining full conditional measures with statistical models that depend on graphoid properties. We intend to report in detail on the consequences of our results for the theory of Bayesian networks in a future publication. We note that future work should develop a theory of Bayesian networks that effectively deals with full conditional measures, either by adopting new factorizations, new concepts of independence, or new separation conditions (perhaps even for the concepts that fail symmetry [53]). Future work should also investigate whether other concepts of independence can be defined so that they satisfy all graphoid properties even in the presence of null events.

In closing, we should appraise the convenience of the layer condition and its use in defining weak coherent and full irrelevance/independence. The conditional Coletti-Scozzafava condition is apparently motivated by a desire to connect epistemic irrelevance with logical independence. Indeed, the Coletti-Scozzafava condition on (B, A) promptly blocks all cases of logical independence in Table 5, except the fifth one, and a symmetrization of the condition blocks all cases. However, the condition is perhaps too strong when directly extended to conditional probabilities, as the conditional Coletti-Scozzafava blocks all logical dependences between the fixed conditioning events and the other relevant variables. We have thus preferred to consider a relaxed condition where logical independence is not

	w_0y_0	w_1y_0	w_0y_1	w_1y_1
x_0	$\lfloor\gamma\alpha\rfloor_1$	$\lfloor\alpha(1-\gamma)\rfloor_1$	$\lfloor(1-\alpha)\gamma\rfloor_1$	$\lfloor(1-\gamma)(1-\alpha)\rfloor_1$
x_1	$\beta\alpha$	$\alpha(1-\beta)$	$(1-\alpha)\beta$	$(1-\beta)(1-\alpha)$

	w_0y_0	w_1y_0	w_0y_1	w_1y_1
x_0	$\alpha\beta$	$\lfloor\alpha\gamma\rfloor_1$	$\lfloor(1-\alpha)\gamma\rfloor_1$	$(1-\alpha)\beta$
x_1	$\alpha(1-\beta)$	$\lfloor\alpha(1-\gamma)\rfloor_1$	$\lfloor(1-\alpha)(1-\gamma)\rfloor_1$	$(1-\alpha)(1-\beta)$

TABLE 8. Failures of versions of contraction ($\alpha,\beta,\gamma \in (0,1)$, with $\alpha \neq \beta \neq \gamma$). The full conditional measure in the top table satisfies (Y EIN X) and (W EIR $X \mid Y$), but it fails (X EIR (W,Y)) (two versions of contraction). The full conditional measure in the bottom table satisfies (X EIN Y) and (X EIR $W \mid Y$), but it fails ((W,Y) EIR X) (two versions of contraction).

always automatically blocked (specifically, in the fifth, sixth and seventh cases in Table 5), and the resulting layer condition does seem to have pleasant properties; for example, it generates a graphoid. Still, one might ask why should the layer condition be adopted at all given that we are not stressing too much the issue of logical dependence. There are two reasons to adopt the layer condition.

First, the layer condition does prevent pathological cases (as depicted in Table 4) that are not related to logical dependence. The characteristic of such examples is that the "relative nullity" of events is not captured just by conditional probability values. We believe that a most promising path in producing a concept of independence for full conditional measures that does satisfy the graphoid properties is to somehow preserve the "distance" in layers for various events as we deal with probabilities. Possibly the machinery of lexicographic probabilities will be required in order to retain information on layer numbers [7, 31]. The layer condition is a step in that direction.

The second reason to adopt the layer condition, now in the particular case of full independence, is that this condition seems to be necessary in order to produce representations of "product" measures that contain some resemblance of factorization. We intend to report on this issue in a future publication.

Finally, there is yet another difference between the layer condition and the Coletti-Scozzafava condition. In the latter the layer numbers are computed with respect to a restricted set of events, whereas in the former we have removed this requirement. Such a difference is again attributed to our desire to keep layer numbers "permanently" attached to events, so that mere marginalization or conditioning does not change these numbers.

Appendix A   Counterexamples to graphoid properties

Tables 2, 3, 7, 8 present violations of decomposition, weak union, contraction and intersection for epistemic irrelevance/independence, coherent irrelevance/independence, h-irrelevance/h-independence and full irrelevance/independence. Table 9 summarizes these counterexamples.

Note that some counterexamples for h-/full irrelevance depend on the fact that X is h-/fully irrelevant to (W, Y) in the top table of Table 7. To verify that this is true, it is necessary to verify the equality

$$P(W, Y | \{X = x\}, A(W, Y)) = P(W, Y | A(W, Y))$$

for $x = \{x_0, x_1\}$ and for every nonempty subset $A(W, Y)$ of $\{w_0 y_0, w_1 y_0, w_0 y_1, w_1 y_1\}$ (there are 15 such subsets).

Direct properties of irrelevance and independence	Epistemic/ Coherent	H-/Full
$(X \perp\!\!\!\perp (W, Y) \,\vert\, Z) \Rightarrow (X \perp\!\!\!\perp W \,\vert\, (Y, Z))$	2	-
$(X \perp\!\!\!\perp Y \,\vert\, Z)$ & $(X \perp\!\!\!\perp W \,\vert\, (Y, Z)) \Rightarrow (X \perp\!\!\!\perp (W, Y) \,\vert\, Z)$	-	2
$(X \perp\!\!\!\perp W \,\vert\, (Y, Z))$ & $(X \perp\!\!\!\perp Y \,\vert\, (W, Z)) \Rightarrow (X \perp\!\!\!\perp (W, Y) \,\vert\, Z)$	3	3

Non-direct/non-reverse properties of irrelevance	Epistemic/ Coherent/ H-/Full
$(X \perp\!\!\!\perp (W, Y) \,\vert\, Z) \Rightarrow (Y \perp\!\!\!\perp X \,\vert\, Z)$	7 (top)
$((W, Y) \perp\!\!\!\perp X \,\vert\, Z) \Rightarrow (X \perp\!\!\!\perp Y \,\vert\, Z)$	7 (bottom)
$(X \perp\!\!\!\perp (W, Y) \,\vert\, Z) \Rightarrow (Y \perp\!\!\!\perp X \,\vert\, (W, Z))$	7 (top)
$((W, Y) \perp\!\!\!\perp X \,\vert\, Z) \Rightarrow (X \perp\!\!\!\perp Y \,\vert\, (W, Z))$	7 (bottom)
$(Y \perp\!\!\!\perp X \,\vert\, Z)$ & $(X \perp\!\!\!\perp W \,\vert\, (Y, Z)) \Rightarrow (X \perp\!\!\!\perp (W, Y) \,\vert\, Z)$	7 (bottom)
$(X \perp\!\!\!\perp Y \,\vert\, Z)$ & $(W \perp\!\!\!\perp X \,\vert\, (Y, Z)) \Rightarrow (X \perp\!\!\!\perp (W, Y) \,\vert\, Z)$	8 (top)
$(Y \perp\!\!\!\perp X \,\vert\, Z)$ & $(W \perp\!\!\!\perp X \,\vert\, (Y, Z)) \Rightarrow (X \perp\!\!\!\perp (W, Y) \,\vert\, Z)$	7 (bottom)
$(X \perp\!\!\!\perp Y \,\vert\, Z)$ & $(X \perp\!\!\!\perp W \,\vert\, (Y, Z)) \Rightarrow ((W, Y) \perp\!\!\!\perp X \,\vert\, Z)$	7 (top)
$(Y \perp\!\!\!\perp X \,\vert\, Z)$ & $(X \perp\!\!\!\perp W \,\vert\, (Y, Z)) \Rightarrow ((W, Y) \perp\!\!\!\perp X \,\vert\, Z)$	8 (bottom)
$(X \perp\!\!\!\perp Y \,\vert\, Z)$ & $(W \perp\!\!\!\perp X \,\vert\, (Y, Z)) \Rightarrow ((W, Y) \perp\!\!\!\perp X \,\vert\, Z)$	7 (top)
$(W \perp\!\!\!\perp X \,\vert\, (Y, Z))$ & $(X \perp\!\!\!\perp Y \,\vert\, (W, Z)) \Rightarrow (X \perp\!\!\!\perp (W, Y) \,\vert\, Z)$	3
$(X \perp\!\!\!\perp W \,\vert\, (Y, Z))$ & $(Y \perp\!\!\!\perp X \,\vert\, (W, Z)) \Rightarrow (X \perp\!\!\!\perp (W, Y) \,\vert\, Z)$	3
$(W \perp\!\!\!\perp X \,\vert\, (Y, Z))$ & $(Y \perp\!\!\!\perp X \,\vert\, (W, Z)) \Rightarrow (X \perp\!\!\!\perp (W, Y) \,\vert\, Z)$	3
$(X \perp\!\!\!\perp W \,\vert\, (Y, Z))$ & $(X \perp\!\!\!\perp Y \,\vert\, (W, Z)) \Rightarrow ((W, Y) \perp\!\!\!\perp X \,\vert\, Z)$	7 (top)
$(W \perp\!\!\!\perp X \,\vert\, (Y, Z))$ & $(X \perp\!\!\!\perp Y \,\vert\, (W, Z)) \Rightarrow ((W, Y) \perp\!\!\!\perp X \,\vert\, Z)$	7 (top)
$(X \perp\!\!\!\perp W \,\vert\, (Y, Z))$ & $(Y \perp\!\!\!\perp X \,\vert\, (W, Z)) \Rightarrow ((W, Y) \perp\!\!\!\perp X \,\vert\, Z)$	7 (top)

TABLE 9.   Summary of counterexamples.   The properties are written using $\perp\!\!\!\perp$; this symbol must be replaced by the concept of interest (epistemic/coherent/h-/full irrelevance/independence). All entries indicate the number of a table containing a counterexample. The top table lists failures of properties for independence and failures of direct properties for irrelevance; the bottom table lists failures of properties of irrelevance that are neither "direct" nor "reverse" versions of graphoid properties. Note that reverse intersection may fail for all concepts in the absence of logical independence.

References

[1] Ernest W. Adams. *A Primer of Probability Logic*. CSLI Publications, 2002.

[2] David Allen and Adnan Darwiche. New Advances in Inference by Recursive Conditioning. In Christopher Meek and Uffe Kjærulff, editors, *UAI '03. Proceedings of the 19th Conference in Uncertainty in Artificial Intelligence. August 7–10, 2003. Acapulco, Mexico*, pages 2–10. Morgan Kaufmann, 2003.

[3] Alexander Balke and Judea Pearl. Counterfactual Probabilities: Computational Methods, Bounds and Applications. In Ramon López de Mántaras and David Poole, editors, *UAI '94: Proceedings of the Tenth Annual Conference on Uncertainty in Artificial Intelligence. July 29–31, 1994. Seattle, Washington, USA*, pages 46–54. Morgan Kaufmann, 1994.

[4] Pierpaolo Battigalli and Pietro Veronesi. A Note on Stochastic Independence without Savage-Null Events. *Journal of Economic Theory*, 70(1):235–248, 1996.

[5] Salem Benferhat, Didier Dubois, and Henri Prade. Nonmonotonic Reasoning, Conditional Objects and Possibility Theory. *Artificial Intelligence*, 92(1–2):259–276, 1997.

[6] Veronica Biazzo, Angelo Gilio, Thomas Lukasiewicz, and Giuseppe Sanfilippo. Probabilistic Logic under Coherence, Model-Theoretic Probabilistic Logic, and Default Reasoning in System P. *Journal of Applied Non-Classical Logics*, 12(2):189–213, 2002.

[7] Lawrence Blume, Adam Brandenburger, and Eddie Dekel. Lexicographic Probabilities and Choice under Uncertainty. *Econometrica*, 59(1):61–79, 1991.

[8] Lawrence Blume, Adam Brandenburger, and Eddie Dekel. Lexicographic Probabilities and Equilibrium Refinements. *Econometrica*, 59(1):81–98, 1991.

[9] Andrea Capotorti, Lucia Galli, and Barbara Vantaggi. How to Use Locally Strong Coherence in an Inferential Process Based on Upper-Lower Probabilities. *Soft Computing*, 7(5):280–287, 2003.

[10] Giulianella Coletti. Coherent Numerical and Ordinal Probabilistic Assessments. *IEEE Transactions on Systems, Man and Cybernetics*, 24(12):1747–1754, 1994.

[11] Giulianella Coletti and Romano Scozzafava. Zero Probabilities in Stochastic Independence. In Bernadette Bouchon-Meunier, Ronald R. Yager, and Lotfi A. Zadeh, editors, *Information, Uncertainty, Fusion*, pages 185–196. Kluwer, 2000.

[12] Giulianella Coletti and Romano Scozzafava. *Probabilistic Logic in a Coherent Setting*, volume 15 of *Trends in Logic*. Kluwer, 2002.

[13] Giulianella Coletti and Romano Scozzafava. Stochastic Independence in a Coherent Setting. *Annals of Mathematics and Artificial Intelligence*, 35(1–4):151–176, 2002.

[14] Robert G. Cowell, Steffen L. Lauritzen, A. Philip David, and David J. Spiegelhalter. *Probabilistic Networks and Expert Systems*. Springer-Verlag, 1999.

[15] Fabio G. Cozman and Peter Walley. Graphoid Properties of Epistemic Irrelevance and Independence. *Annals of Mathematics and Artificial Intelligence*, 45(1–2):173–195, 2005.

[16] Lucio Crisma. The notion of stochastic independence in the theory of coherent probability. Technical Report 1/99, Quaderni di Dipartimento di Matematica Applicata alle Scienze Economiche, Statistiche e Attuariali "Bruno de Finetti", 1999.

[17] Adnan Darwiche and Moises Goldszmidt. On the Relation between Kappa Calculus and Probabilistic Reasoning. In Ramon López de Mántaras and David Poole, editors, *UAI '94: Proceedings of the Tenth Annual Conference on Uncertainty in Artificial Intelligence. July 29–31, 1994. Seattle, Washington, USA*, pages 145–153. Morgan Kaufmann, 1994.

[18] A. Philip Dawid. Conditional Independence in Statistical Theory. *Journal of the Royal Statistical Society. Series B (Methodological)*, 41(1):1–31, 1979.

[19] A. Philip Dawid. Conditional Independence. In Samuel Kotz, Campbell B. Read, and David L. Banks, editors, *Encyclopedia of Statistical Sciences*, pages 146–153. Wiley, 1999. Volume 2.

[20] Bruno de Finetti. *Theory of Probability*. Wiley, 1974.

[21] Rina Dechter and Robert Mateescu. Mixtures of Deterministic-Probabilistic Networks and their AND/OR Search Space. In Max Chickering and Joseph Halpern, editors, *UAI '04. Proceedings of the 20th Conference in Uncertainty in Artificial Intelligence. July 7-11 2004. Banff, Canada*, pages 120–129. AUAI Press, 2004.

[22] Morris H. DeGroot. *Probability and Statistics*. Addison-Wesley, 1986.

[23] Lester E. Dubins. Finitely Additive Conditional Probability, Conglomerability and Disintegrations. *Annals of Statistics*, 3(1):89–99, 1975.

[24] David Galles and Judea Pearl. Axioms of Causal Relevance. *Artificial Intelligence*, 97(1–2):97–1, 1996.

[25] Hector Geffner. *Default Reasoning: Causal and Conditional Theories*. MIT Press, 1992.

[26] Hector Geffner. High Probabilities, Model Preference and Default Arguments. *Mind and Machines*, 2:51–70, 1992.

[27] Dan Geiger, Thomas Verma, and Judea Pearl. Identifying Independence in Bayesian Networks. *Networks*, 20:507–534, 1990.

[28] Moisés Goldszmidt and Judea Pearl. Qualitative Probabilities for Default Reasoning, Belief Revision, and Causal Modeling. *Artificial Intelligence*, 84(1–2):57–112, 1996.

[29] Alan Hájek. What Conditional Probability Could Not Be. *Synthese*, 137:273–323, 2003.

[30] Joseph Y. Halpern. Lexicographic Probability, Conditional Probability, and Nonstandard Probability. In Johan van Benthem, editor, *Proceedings of the 8th Conference on Theoretical Aspects of Rationality and Knowledge (TARK-2001). Certosa di Pontignano, University of Siena, Italy. July 8–10, 2001*, pages 17–30, 2000.

[31] Peter J. Hammond. Elementary Non-Archimedean Representations of Probability for Decision Theory and Games. In Paul Humphreys, editor, *Patrick Suppes: Scientific Philosopher; Volume 1*, pages 25–59. Kluwer, 1994.

[32] Peter J. Hammond. Consequentialism, non-Archimedean Probabilities, and Lexicographic Expected Utility. In Cristina Bicchieri, Richard Jeffrey, and Brian Skyrms, editors, *The Logic of Strategy*, pages 39–66. Oxford University Press, 1999.

[33] Peter J. Hammond. Non-Archimedean Subjective Probabilities in Decision Theory and Games. *Mathematical Social Sciences*, 38:139–156, 1999.

[34] Silvano Holzer. On Coherence and Conditional Prevision. *Bollettino Unione Matematica Italiana Serie VI*, 1:441–460, 1985.

[35] John M. Keynes. *A Treatise on Probability*. Macmillan and Co., 1921.

[36] Sarit Kraus, Daniel Lehmann, and Menachem Magidor. Nonmonotonic Reasoning, Preferential Models and Cumulative Logics. *Artificial Intelligence*, 14(1):167–207, 1990.

[37] Peter Krauss. Representation of Conditional Probability Measures on Boolean Algebras. *Acta Mathematica Academiae Scientiarum Hungaricae*, 19(3–4):229–241, 1968.

[38] Isaac Levi. *The Enterprise of Knowledge*. MIT Press, 1980.

[39] Ernesto San Martin, Michel Mouchart, and Jean-Marie Rolin. Ignorable Common Information, Null Sets and Basu's First Theorem. *Sankya; the Indian Journal of Statistics*, 67:674–698, 2005.

[40] František Matúš and Milan Studený, editors. *Workshop on Conditional Independence Structures and Graphical Models. Book of Abstracts*. ÚTIA AV CR, 1999.

[41] Vann McGee. Learning the Impossible. In Ellery Bells and Brian Skyrms, editors, *Probability and Conditionals: Belief Revision and Rational Decision*, pages 179–199. Cambridge University Press, 1994.

[42] Roger B. Myerson. *Game Theory: Analysis of Conflict*. Harvard University Press, 1997.

[43] Judea Pearl. *Probabilistic Reasoning in Intelligent Systems: Networks of Plausible Inference*. Morgan Kaufmann, 1988.

[44] Judea Pearl. Probabilistic Semantics for Nonmonotonic Reasoning: A Survey. In Ronald J. Brachman, Hector J. Levesque, and Raymond Reiter, editors, *Proceedings of the 1st International Conference on Principles of Knowledge Representation and Reasoning (KR'89). Toronto, Canada, May 15–18, 1989*, pages 505–516. Morgan Kaufmann, 1989.

[45] Karl R. Popper. *The Logic of Scientific Discovery*. Hutchinson, 1975.

[46] Eugenio Regazzini. Finitely Additive Conditional Probability. *Rendiconti del Seminario Matematico e Fisico*, 55:69–89, 1985.

[47] Alfred Renyi. On a New Axiomatic Theory of Probability. *Acta Mathematica Hungarica*, 6:285–335, 1955.

[48] Teddy Seidenfeld, Mark J. Schervish, and Joseph B. Kadane. Improper Regular Conditional Distributions. *Annals of Probability*, 29(4):1612–1624, 2001.

[49] Wolfgang Spohn. Ordinal Conditional Functions: A Dynamic Theory of Epistemic States. In William L. Harper and Brian Skyrms, editors, *Causation in Decision, Belief Change and Statistics*, volume II, pages 105–134. Springer, 1988.

[50] Robert Stalnaker. Probability and Conditionals. *Philosophy of Science*, 37:64–80, 1970.

[51] Bas C. van Fraassen. Fine-Grained Opinion, Probability, and the Logic of Full Belief. *Journal of Philosophical Logic*, 24:349–377, 1995.

[52] Barbara Vantaggi. Conditional Independence in A Coherent Finite Setting. *Annals of Mathematics and Artificial Intelligence*, 32(1-4):287–313, 2001.

[53] Barbara Vantaggi. Graphical Representation of Asymmetric Graphoid Structures. In Jean-Marc Bernard, Teddy Seidenfeld, and Marco Zaffalon, editors, *ISIPTA '03. Proceedings of the Third International Symposium on Imprecise Probabilities and Their Applications held in Lugano, Switzerland*, pages 560–574. Carleton Scientific, 2003.

[54] Barbara Vantaggi. Qualitative Bayesian Networks with Logical Constraints. In Thomas D. Nielsen and Nevin Lianwen Zhang, editors, *Symbolic and Quantitative Approaches to Reasoning with Uncertainty, 7th European Cnference, ECSQARU 2003, Aalborg, Denmark, July 2-5, 2003. Proceedings*, volume 2711 of *Lecture Notes in Computer Science*, pages 100–112. Springer, 2003.

[55] Jiřina Vejnarová. Conditional Independence Relations in Possibility Theory. In Gert De Cooman, Fabio Gagliardi Cozman, Serafín Moral, and Peter Walley, editors, *ISIPTA '99. Proceedings of the First International Symposium on Imprecise Probabilities and Their Applications, Held at the Conference Center "Het Pand" of the Universiteit Gent, Ghent, Belgium, 29 June – 2 July 1999*, pages 343–351, 1999.

[56] Peter Walley. *Statistical Reasoning with Imprecise Probabilities*. Chapman and Hall, 1991.

[57] Peter Walley, Renato Pelessoni, and Paolo Vicig. Direct Algorithms for Checking Consistency and Making Inferences from Conditional Probability Assessments. *Journal of Statistical Planning and Inference*, 126(1):119–151, 2004.

[58] Peter M. Williams. Coherence, Strict Coherence and Zero Probabilities. In Robert E. Butts and Jaakko Hintikka, editors, *Contributed Papers to*

the Fifth International Congress of Logic, Methodology and Philosophy of Science, pages III.29–III.30. University of Western Ontario Press, 1975.

[59] Nic Wilson. An Order of Magnitude Calculus. In Philippe Besnard and Steve Hanks, editors, *Proceedings of the 11th Conference on Uncertainty in Artificial Intelligence (1995). Aug 18–20 1995, Montreal, QC*, pages 548–555. Morgan Kaufmann, 1995.

Benedikt **Löwe**, Eric **Pacuit**, Jan-Willem **Romeijn** (*eds.*)
Foundations of the Formal Sciences VI
Reasoning about Probabilities and Probabilistic Reasoning

Indeterminacy and Choice

Jeffrey Helzner*

Department of Philosophy, Columbia University, 708 Philosophy Hall, 1150 Amsterdam
Ave, New York NY, 10027, United States of America
E-mail: jh2239@columbia.edu

1

L. J. Savage's subjective expected utility theory, as presented in [25], is viewed by some as the final word on normative accounts of decision making, but it is instructive to recall how the additive model that underlies the principle of maximizing expectation has been modified over time in order to encompass a wider class of decision situations. In keeping with what is now a well-known classification, we may distinguish between two basic types of decision situations, *decision making under risk* and *decision making under uncertainty* [20]. The distinction between these two is that in an instance of the former the decision maker has access to an objective probability distribution over the possible states, while in the latter there is no such access.[1]

Parlor games and lotteries are instances of decision making under risk. Indeed, much of the early work in this area focused on gambling situations. The recommendation to maximize expected value, the origins of which date at least as far back as 1662 with the publication of Arnaud's *Port-Royal Logic* [10], requires the rational agent to select an alternative that maximizes the following index:

$$\text{EV}(a) = \sum_{i \in S} \mathbf{p}(i) a(i) \tag{1.1}$$

where S is a set of mutually exclusive and exhaustive states, a is an act,

---

*I would like to thank Horacio Arlo-Costa, Isaac Levi, and Teddy Seidenfeld for many
years of stimulating conversation concerning the topics that are considered in this paper.
[1]While Luce and Raiffa's classification continues to enjoy popularity within the decision sciences, it is worth noting that some, e.g., [19], have expressed doubts about their taxonomy.

Received by the editors: 10 September 2007; 25 February 2008.
Accepted for publication: 8 April 2008.

i.e., a function from the states of S to a set of consequences, and $\mathbf{p}$ is a probability distribution on S representing the objective probability that is assumed to be accessible to the agent in instances of decision making under risk.

It is implicit in the formulation of (1.1) that expected value is restricted to acts with monetary consequences or something similar that supports the arithmetic operations of (1.1). However, this is not the main critique of expected value. The main critique of expected value appears to originate with D. Bernoulli's 1738 recognition of the diminishing marginal utility of wealth as a resolution of the St. Petersburg paradox. With this recognition, Bernoulli had arrived at the core of the expected utility thesis, which holds that the rational agent should select an alternative that maximizes the following index:

$$\mathrm{EU}(a) = \sum_{i \in S} \mathbf{p}(i) u_i(a(i)) \tag{1.2}$$

where u_i represents the decision maker's utility judgments regarding the consequences that are associated with state i, while S, a, and $\mathbf{p}$ remain as described in connection with (1.1).

Unlike expected value, expected utility appears to be applicable in all instances of decision making under risk, at least insofar as the relevant consequences support a real-valued notion of utility. On the face of it, the major restriction of (1.2) seems to be the very feature that distinguishes decision making under risk, namely the agent's access to an objective probability over the possible states. This restriction would seem to invite doubts concerning the significance of such a normative thesis, since objective probabilities fail to be salient in many real-world decisions. In response to such doubts, those who are enamored with the recommendation to maximize expectation may appeal to subjective expected utility theory [SEU], according to which the rational agent is to select an act that maximizes the following index:

$$\mathrm{SEU}(a) = \sum_{i \in S} p(i) u_i(a(i)) \tag{1.3}$$

where S, a, and u_i retain their interpretation from (1.2), but, unlike (1.1) and (1.2), the probability p employed in (1.3) is a subjective probability distribution, which, in keeping with the subjectivist tradition of [24], is to be interpreted as representing the decision maker's degrees of belief concerning the various possible states.

2

According to subjectivists like Savage, (1.3) is applicable to all instances of decision making under uncertainty, since the rational agent will always have a subjective probability over the possible states. In light of this, and the structural similarity between (1.2) and (1.3), it has been said that the classical subjectivists reduce uncertainty to risk [20]. While this reduction is regarded as unproblematic in some quarters, others suspect that the indicated reduction —and, more generally, subjective expected utility theory's treatment of uncertainty— are grounds for significant concern about the normative status of the SEU thesis. Nowhere has this concern been more apparent than in connection with the work of D. Ellsberg as reported in [3, 4]. In following a line that may traced back to earlier work by J. M. Keynes [13] and F. H. Knight [14], Ellsberg argued that there exist "uncertainties that are not risks". At the heart of Ellsberg's argument is a pair of unusually compelling examples, the most famous of which is as follows:

Example 1 (Ellsberg [3]). A ball is to be selected at random from an urn containing 90 balls. 30 of those balls are red. Each of the remaining 60 balls in the urn is either black or white, although the exact ratio of black balls to white balls is unknown. Consider the following two decision problems.

Problem 1

	Red	Black	White
I	$100	$0	$0
II	$0	$100	$0

Problem 2

	Red	Black	White
III	$100	$0	$100
IV	$0	$100	$100

In connection with this example, which has come to be known as *Ellsberg's three-color problem*, it has been observed that a significant number of people select *I* as uniquely admissible in Problem 1 and *IV* as uniquely admissible in Problem 2. It is clear that such a choice pattern is incompatible with subjective expected utility theory. More generally, this choice pattern is incompatible with every theory in which preference satisfies independence and determines admissibility in the usual way; i.e., an available alternative is admissible as long as it is not strictly dispreferred to some other available alternative.[2]

---

[2]In this context, *independence* requires that, for all acts f and g, those states where f and g agree have no bearing on preferences between these acts; that is, the preference would be maintained if f and g were modified uniformly at those states where they agree. In Problem 1 above, acts *I* and *II* agree on White. Whatever the preference

There are several different views concerning the significance of this incompatibility. Some people have interpreted the incompatibility as additional evidence that SEU is not descriptively adequate. Inspired by this interpretation, some decision scientists have sought to preserve the normative status of SEU while they look for a distinct, descriptive theory that can explain the observed deviations from SEU. A significant amount of the research in this direction has followed the so-called "heuristics and biases" paradigm this is most clearly associated with D. Kahneman and A. Tversky [29]. Theories within this tradition have tended to attribute deviations from rationality (as explicated in SEU or some similar normative thesis) to the presence of certain psychological effects. Fox and Tversky's "comparative ignorance hypothesis" [6] is a notable example of work done within this tradition that is explicit in its claims of being able to explain the particular deviations which have been observed in connection with Ellsberg's examples.

In contrast to what has just been described, there is a second interpretation that takes those deviations that have been discussed in connection with Ellsberg's examples as motivation for a normative alternative to SEU. Those who have taken this interpretation are moved by the fact that the indicated deviations are often exhibited by subjects who are aware of the requirements of SEU as well as their failure to comply with these requirements. Those who would glibly answer such an interpretation by suggesting that normative theses need not be responsive to empirical matters, such as observed deviations, should be prepared to explain how a normative thesis is to be grounded without an appeal, delayed as it may be, that rests ultimately on a pretheoretical judgment concerning what counts as rational.

In his analysis of the deviations from SEU that he observed in connection with the three-color problem, Ellsberg, who in [3] clearly adopts the second of these interpretations, suggested a decision rule that is informed by both a concern for security, which had figured prominently in game-theoretic studies of statistical methods, and the practice of taking expectations, which, by that time, had long since achieved its central position in decision theory. This decision rule —which is sometimes referred to as the "Restricted Bayes-Hurwicz Criterion"— may be specified in terms of three parameters: a set Y of probability distributions over the relevant set of states, a distinguished element $y_0 \in Y$, and a value $\lambda \in [0, 1]$. According to the intended interpretation, Y consists of all those distributions that are consistent with the decision maker's information, while y_0 represents the decision maker's

---

is with respect to I and II, independence requires that it is maintained in Problem 2, since the acts in Problem 2 are obtained from I and II via a uniform modification of the indicated sort. Independence, as it is known in the general study of additive models, is discussed at length in [15, 30, 12] and within the context of the "sure-thing principle" in [25].

"estimated" distribution. Let $E_p^u(a)$ denote the expected utility of act a against utility u and probability distribution p. Given a set X of probabilities, let $S_X(a)$ denote the infimum of $\{E_p^u(a) \mid p \in X\}$; here, $S_X(a)$ is supposed to be interpreted as the security of a with respect to X and u. The Restricted Bayes-Hurwicz Criterion recommends that the decision maker select an act that maximizes the following index:

$$\lambda E_{y_0}(x) + (1 - \lambda)S_Y(x) \qquad (2.1)$$

where here λ is to be interpreted as a measure of the decision maker's willingness to make tradeoffs between expectation and security. Returning to Ellsberg's three-color problem (Example 1), suppose that

$$Y = \left\{ p \mid p(\text{Red}) = \frac{1}{3} \right\}$$

and

$$y_0(\text{Red}) = y_0(\text{Black}) = y_0(\text{White})$$

and that u is strictly increasing as a function of dollars, then it easy to verify that, for $\lambda < \frac{1}{2}$, I is uniquely admissible in Problem 1 while IV is uniquely admissible in Problem 2.

In contrast to (2.1), which uses a weighted average in order to integrate expected utility considerations and security considerations, Isaac Levi has advocated a two-stage, lexicographic decision rule as a normative replacement of SEU [17, 19]. As before, Y is a the set of all distributions that are consistent with the decision maker's information. If A is a set of alternatives, then $a \in A$ is *E-admissible* in A just in case a maximizes expected utility over A for at least one distribution in Y. Security considerations enter at the second stage so that $a \in A$ is admissible in A just in case a is E-admissible in A and a has maximal security among the E-admissible alternatives in A.

Levi has a fairly liberal view regarding what may serve as a security index. For example, he allows for vacuous security, in which case admissibility coincides with E-admissibility, as well as the nontrivial security index that was discussed in conjunction with the Restricted Bayes-Hurwicz Criterion. By adopting this second security index and applying the resulting instance of Levi's criterion to Ellsberg's three-color problem, again with $Y = \{p \mid p(\text{Red}) = \frac{1}{3}\}$, it easy to verify that I is uniquely admissible in Problem 1 while IV is uniquely admissible in Problem 2.

3

While each of the decision rules discussed in the previous section may be regarded as a generalization of SEU, where agreement with SEU obtains when

Y is a singleton and security is vacuous, these rules differ in important ways. The decision rule favored by Ellsberg results in a real-valued index and thus produces an ordering on the set of alternatives. Of course, since this rule supports the I/IV pattern that was discussed in connection with the three-color problem, it is clear that this ordering need not satisfy independence. Additional work based on the idea of maintaining ordering while relaxing independence has developed along several lines, e.g., philosophically in [7] and axiomatically in [8]. Alternatively, the criterion advocated by Levi results in a notion of admissibility that, in general, cannot be reduced to an ordering. To see that this is so we recall two well-known properties that were proposed by A.K. Sen [27, 16].

If X is a nonempty set of alternatives, then we will write $\mathcal{X}$ for the set of all finite, nonempty subsets of X. C is a *choice function* on X just in case $C : \mathcal{X} \to \mathcal{X}$ satisfies $C(Y) \subseteq Y$ for all $Y \in \mathcal{X}$.[3] $C(Y)$ will be interpreted as the set of alternatives that the agent would judge as admissible if it were presented with menu Y. The requirement that C is a function on $\mathcal{X}$ guarantees that each nonempty menu has at least one admissible alternative. The requirement $C(Y) \subseteq Y$ guarantees that the admissible alternatives are among those that are available. With these preliminaries out of the way we may now recall Sen's α and β properties.

$$\text{For all } S, T \in \mathcal{X} \text{ such that } S \subseteq T, \text{ if } x \in S \cap C(T), \text{ then } \qquad (\alpha)$$
$$x \in C(S).$$

$$\text{For all } S, T \in \mathcal{X} \text{ such that } S \subseteq T, \text{ if } x, y \in C(S) \text{ and } y \in \qquad (\beta)$$
$$C(T), \text{ then } x \in C(T).$$

The idea behind these conditions is easy to grasp. Property α says that removing some of the competition does not result in an admissible alternative being demoted to an inadmissible alternative. Property β says that if two alternatives are admissible with respect to a given menu, then, as far as admissibility goes, these two alternatives cannot be distinguished from each other by enlarging the menu that was given. Sen's α and β properties are well-known to be jointly necessary and sufficient for reducing admissibility to an ordering.[4] Levi's criterion permits violations of these conditions and

---

[3]The study of set-valued choice functions is well-established in the microeconomics literature. The reader may wish to consult [27, 16] for a review of basic results connected with the study of such functions.

[4]We use the term 'ordering' on X to refer to those binary relations on X that are asymmetric and negatively transitive, i.e., those binary relations that satisfy the following conditions for all $x, y, z \in X$: (1) if $x \succ y$, then not $y \succ x$ and (2) if neither $x \succ y$ nor $y \succ z$, then not $x \succ z$. Sen's α and β are well-known to be jointly necessary and sufficient for the existence of an ordering $\succ$ on X such that, for all $Y \in \mathcal{X}$, $x \in C(Y)$ iff $x \in Y$ and there is no $y \in Y$ such that $y \succ x$.

thus, in general, does not provide for such a reduction. The following example illustrates the manner in which Levi's criterion can lead to violations of both conditions.

Example 2 (Levi [17]). Suppose that a random selection is to be taken from the urn that is described in Example 1. Consider the following three acts:

	Red	Black	White
e	3	0	3
f	3	3	0
g	$\frac{3}{2}$	$\frac{3}{2}$	$\frac{3}{2}$

Assume that $Y = \{p \mid p(\text{Red}) = \frac{1}{3}\}$ and u is linear in dollars. Let C be the choice function that is determined by Levi's admissibility criterion as applied to the non-vacuous security index that was discussed above. It is easy to verify that $C(\{e, f, g\}) = \{e, f\} = C(\{e, f\})$ while $C(\{e, g\}) = \{g\} = C(\{f, g\})$ and thus C violates α. Furthermore, if C' is the choice function that is determined by Levi's admissibility criterion as applied to vacuous security, then it is clear that $C'(\{e, f, g\}) = \{e, f\} = C'(\{e, f\})$ while $C'(\{e, g\}) = \{e, g\}$ and $C'(\{f, g\}) = \{f, g\}$ and thus C' violates β.

4

In our discussion of uncertainty we have been focusing on the challenges that this sort of indeterminacy can introduce with respect to the ordering assumptions of expected utility models. It is not difficult to imagine analogous difficulties within other contexts. For example, imagine a multiattribute choice problem where your assessment of the relative importance of the attributes is indeterminate.[5] This could happen in more than one way. One way that this could happen is if the agent's assessment of the attribute weights is conditional on some proposition, where the truth-value of this proposition eludes the agent. If the agent has a determinate probability with respect to this proposition, then the agent might be able to use this probability as the basis for combining the conditional assessments into an unconditional assessment. However, if the agent's credal state is indeterminate with respect to this proposition, then this proposal would seem to result in a family of attribute weights, indexed by a set of probabilities. That is, this proposal to obtain an unconditional weight by averaging the conditional weights presupposes a numerically precise probability. Alternatively, instances of unresolved value conflict, such as those considered in [18],

---

[5]Problems involving 'attribute weight uncertainty' have received some attention in the study of consumer behavior, e.g., in [31] and [11], while [9] considers difficulties concerning the status of the ordering assumption for multiattribute models when there is indeterminacy with respect to attribute weights.

might result in competing assessments regarding the relative importance of
the attributes involved.

Whether indeterminacy with respect to attribute weight derives from
uncertainty in belief or conflict in value, this sort of indeterminacy has not
received nearly as much attention as its counterpart in the usual act-state
formulation of decision making under uncertainty. Moreover, studies of
this sort of indeterminacy within the context of descriptive models, e.g.,
[11], have tended to endorse the traditional ordering assumptions. Taking
other modes of accommodation seriously, e.g., something along the lines
of Levi's lexicographic criterion, would seem to require a reexamination of
the manner in which such descriptive models are tested. For example, it
seems that taking such an account seriously would necessitate abandoning
the standard reduction to pairwise choice, i.e., choice from menus with just
two elements, that is supported by the joint satisfaction of Sen's α and β
properties.[6] Thus, in taking such an alternative account seriously we are
confronted with the problem of finding relaxations of these properties that
could be tested in connection with such an account. The taxonomy of choice
properties outlined in [28] offers a starting point from which to address
this sort of problem. One of the first relaxations that Sen considers is the
following weakened version of α, which is attributed to P.C. Fishburn [5]:

Fishburn's Condition A5 If $Y \in \mathcal{X}$ has more than two elements and
$y \in C(Y)$, then $y \in C(\{y, z\})$ for some $z \in Y$ such that $y \neq z$.

Proposition 4.1. Fishburn's A5 is satisfied by every choice function that
is based on the Levi criterion considered above.

Indeed, we can verify this condition for the following generalization of
Levi's criterion. First note that given a set $\mathcal{O}$ of orderings on X (i.e., asymmetric and negatively transitive binary relations on X) we can formulate
first-tier admissibility as follows: $y \in Y$ is *first-tier admissible* in Y iff there
is at least one $\succ$ in $\mathcal{O}$ according to which there is no $z \in Y$ such that $z \succ y$.
If $\succ_S$ is an ordering on X, the intended interpretation of which is some
notion of security, then, in keeping with the earlier presentation of Levi's
criterion, we can formulate admissibility as follows: $y \in Y$ is *admissible* in Y
iff it is first-tier admissible in Y and there is no z that is first-tier admissible

---

[6]It is well known that if C satisfies α and β, then admissibility according to C is
representable as maximization of the relation R that is given as follows: xRy iff $x \in C(\{x,y\})$. Hence, to reconstruct C it is sufficient to restrict attention to pairwise choice
[27].

in Y such that $z \succ_S y$.[7]

Proof of Proposition 4.1. Assume that C is a choice function on X for which there exists $\mathcal{O}$ and $\succ_S$, as above, such that, for all $Y \in \mathcal{X}$, $y \in C(Y)$ iff y is admissible in Y. Suppose that Y is a subset of X that contains more than two elements. Assume that $y_0 \in C(Y)$. *Case 1*: Suppose that $C(Y)$ contains an element y_1 where $y_1 \neq y_0$. Since y_0 is first-tier admissible in Y it follows that y_0 is first-tier admissible in $\{y_0, y_1\}$. As $y_0, y_1 \in C(Y)$, it follows that y_0 and y_1 have the same security level, i.e., neither $y_0 \succ_S y_1$ nor $y_1 \succ_S y_0$. Hence, $y_0 \in C(\{y_0, y_1\})$. *Case 2*: Suppose that $C(Y) = \{y_0\}$. If y_0 is the only element that is first-tier admissible in Y, then y_0 is the only element that is first-tier admissible in $\{y_0, y_1\}$, where y_1 is any element of Y that is distinct from y_0. Thus, $y_0 \in C(\{y_0, y_1\})$. On the other hand, if y_1 is distinct from y_0 and first-tier admissible in Y, then $y_0 \succ_S y_1$. Since y_0 is first-tier admissible in Y, it follows that it is first-tier admissible in $\{y_0, y_1\}$. Finally, since $y_0 \succ_S y_1$, it follows that $y_0 \in C(\{y_0, y_1\})$.   Q.E.D.

Sen also considers the following condition as a weakening of α, although along lines different than that of Fishburn's A5:

If $Y \in \mathcal{X}$, then there is some $x \in C(Y)$ such that $x \in C(\{x, y\})$ for all $y \in Y$. $(\alpha(-))$

A still weaker version of α, also considered in [28], restricts $\alpha(-)$ to triples. With respect to the suggested weakening Sen writes as follows:

> Property $\alpha(-)$ may be weakened even further if we demand its fulfillment only for the smallest subsets larger than a pair (for pairs, of course, the problem of contraction consistency does not arise), viz., for triples. This we call the "weakest contraction-consistency" property.

If Y is a three-element subset, then there is some $x \in C(Y)$ such that $x \in C(\{x, y\})$ for all $y \in Y$. $(\alpha(--))$

Unlike the weakened version of α represented by Fishburn's A5, choice functions based on Levi's criterion need not satisfy $\alpha(--)$; thus, such choice functions need not satisfy the stronger $\alpha(-)$. Consider the choice function C from Example 2. As neither e nor f are admissible in a pairwise contest against g, it follows that C fails to satisfy these weakened versions of α.

[7]While retaining some of the structural features of Levi's decision theory, e.g., lexicographic integration of first-tier and second-tier admissibility, the suggested generalization does not maintain certain conditions, e.g., convexity of the set of first-tier orderings, that are required by Levi.

Though not discussed in the context of the taxonomy presented in [28], it is worth noting that all choice functions within the suggested generalization satisfy the following weakening of α:

$$\text{If } Y, Z \in \mathcal{X} \text{ and } Y \subseteq C(Z), \text{ then } Y \subseteq C(Y). \qquad (\mathcal{A})$$

It is clear that α implies $\mathcal{A}$: Assume α and suppose that $Y \subseteq C(Z)$. Since $Y \subseteq C(Z) \subseteq Z$, it follows from α that $Y \cap C(Z) \subseteq C(Y)$, but $Y = Y \cap C(Z)$. Also, as noted, the following proposition holds.

Proposition 4.2. Condition $\mathcal{A}$ is satisfied by every choice function that is based on the Levi criterion considered above.

Proof. Suppose that C is a choice function of the indicated sort. By assumption, first-tier admissibility for C is given by some set $\mathcal{O}$ of orderings, and second-tier admissibility is given by some ordering $\succ_S$ that is intended to capture some notion of security. Assume that $Y \subseteq C(Z)$. If $y \in Y$, then, since $y \in C(Z)$, y is maximal in Z with respect to some ordering in $\mathcal{O}$. Since $Y \subseteq Z$, it follows that y is maximal in Y with respect to this same ordering in $\mathcal{O}$. Hence, y is first-tier admissible in Y. If y is not second-tier admissible in Y, then there is a $z \in Y$ such that $z \succ_S y$, but this contradicts the assumption that $y \in C(Z)$, since it would have been eliminated from contention by having a security level inferior to that of z. It follows that y is also second-tier admissible in Y. Q.E.D.

Also discussed by Sen in [28] are various weakened versions of β, the strongest of which is following condition:

$$\text{If } Y, Z \in \mathcal{X} \text{ and } Y \subseteq Z, \text{ then it is not the case that } C(Z) \subset C(Y). \qquad (\varepsilon)$$

It is clear that this condition is implied by β. Also, it is easy to see that the given generalization of Levi's criterion, which was shown above to satisfy Fishburn's A5, need not satisfy ε. Let $Y = \{a, b, c\}$ and $Z = \{a, b\}$. Suppose that $\mathcal{O}$ contains the following orderings

$$a \succ_1 b \succ_1 c \text{ and } c \succ_2 b \succ_2 a$$

and suppose that security is given by

$$a \sim b \succ_S c$$

where $a \sim b$ means that a and b are equivalent with respect to security (i.e., neither $a \succ_S b$ nor $b \succ_S a$). If C is the associated choice function, then $C(\{a, b, c\}) = \{a\}$ while $C(\{a, b\}) = \{a, b\}$.

Finally, the following still further weakening of β is considered in [27, 28]:

$$\text{If } Y, Z \in \mathcal{X} \text{ and } Y \subseteq Z \text{ and } x, y \in C(Y), \text{ then } C(Z) \neq \{x\}. \qquad (\delta)$$

This is the most tolerant weakening of β that is considered in [28]. It is clear that the example that was used to show the possibility of ε-violations also shows that the given generalization of Levi's criterion need not satisfy δ. It is not difficult to find a weakening of β that is satisfied by the given generalization of Levi's criterion. For example, consider the following condition:

> For all $S, T \in \mathcal{X}$ such that $S \subseteq T$, if $x, y \in C(S)$ and, for all $z \in T - \{x, y\}$, $z \notin C(\{x, z\})$ and $C(T)$ contains at least two $(\beta(-))$
> elements, then $x \in C(T)$ implies $y \in C(T)$.

Proposition 4.3. The above generalization of Levi's criterion satisfies $\beta(-)$.

Proof. Suppose that $S, T \in \mathcal{X}$ such that $S \subseteq T$, $x, y \in C(S)$ and, for all $z \in T - \{x, y\}$, $z \notin C(\{x, z\})$ and $C(T)$ has at least two elements. Assume that $x \in C(T)$. It is enough to show that there is no $z \in T - \{x, y\}$ such that $z \in C(T)$. Suppose that $z \in T - \{x, y\}$. By assumption, $z \notin C(\{x, z\})$. Either z is not first-tier admissible in $\{x, z\}$ or z is not second-tier admissible in $\{x, z\}$ (i.e., z has lower security than x). If z is not first-tier admissible in $\{x, z\}$, then z cannot be first-tier admissible in T, since it is dominated by x under each permissible ranking. If z is first-tier admissible in $\{x, z\}$, then it must be the case that z has lower security than x. Since z has lower security than x and $x \in C(T)$, it follows that z is not an element of $C(T)$.   Q.E.D.

Before turning to the rather different approach to characterization that is considered in the next section, it is worth recalling Sen's discussion of "normal" choice functions, a class that includes those satisfying both α and β as a special case. Given a choice function C on $\mathcal{X}$, let R_C be the binary relation on X that is defined as follows for all $x, y \in X$: xR_Cy iff $x \in C(Y)$ for some $Y \in \mathcal{X}$ such that $x, y \in Y$. C is *normal* just in case $C(Y) = \{y \in Y \mid yR_Cz \text{ for all } z \in Y\}$ for all $Y \in \mathcal{X}$. As discussed in [28], C is normal if and only if it satisfies α along with the following condition:

> For all $\{Y_i\}_{i \in I}$ such that $Y_i \in \mathcal{X}$ for all $i \in I$, if $x \in C(Y_i)$
> for all $i \in I$, then $C(\bigcup_{i \in I} Y_i)$. (γ)

Also noted by Sen is the fact the choice functions that satisfy both α and β are the normal choice functions that are generated by an ordering. Mere satisfaction of α and γ does not preclude the possibility of being generated by an odd preference relation. For example, if C is the choice function on $\{a, b, c\}$ that is defined by $C(\{a, b, c\}) = \{a\}$, $C(\{a, b\}) = \{a\}$, $C(\{b, c\}) =$

$\{b\}$ and $C(\{a,c\}) = \{a,c\}$, then C satisfies α and γ but not β. Note that in this example $bR_C c$ and $cR_C a$, but not $bR_C a$. It is not hard to see that lexicographic choice functions of the sort considered by Levi need not be normal. For example, neither of the choice functions in Example 2 satisfy γ. In fact, choice functions of this sort do not even need to satisfy the following restricted versions of α and γ:

For all $Y \in \mathcal{X}$, if $x \in C(Y)$, then $x \in C(\{x,y\})$ for all $y \in Y$.   (α2)

For all $Y \in \mathcal{X}$, if $x \in C(\{x,y\})$ for all $y \in Y$, then $x \in C(Y)$.   (γ2)

Joint satisfaction of α2 and γ2 can be taken as an alternative characterization of the normal choice functions.[8] Thus, in light of the previous remarks, it is clear that choice functions of the sort suggested by Levi need not satisfy both of these conditions. In fact, as C in Example 2 shows, choice functions of this sort can violate both α2 and γ2.

5

In the previous section we considered a sort of generalization of the choice functions considered by Levi. The choice functions on X that satisfy this generalization are those that can be specified by a pair $\langle \mathcal{O}, \succ_S \rangle$, where $\mathcal{O}$ is a set of orderings on X that determines a notion of first-tier admissibility and $\succ_S$ is an ordering on X that supplies a notion of "security" that is applied as a second-tier consideration. The structure of these functions appears to be less transparent than those of more familiar classes, e.g., choice functions that are generated by a single ordering on the set of alternatives. The identification of different characterizations might lead to a better understanding of this seemingly less transparent class. There are some results of this sort. For example, Seidenfeld, Schervish, and Kadane characterize the class of "coherent" choice functions on a space of lotteries, where first-tier admissibility for a coherent choice function is given by a set of orderings, each having an expected utility representation, and second-tier admissibility is vacuous [26]. Important examples of work in this area concerning the general theory of choice functions is provided by Aizerman and Malishevski in [2, 1] and Nehring and Puppe [22, 23], which offer characterizations for a subclass of the generalization under consideration, namely those on a finite set that employ a vacuous security rule in the second tier.

Thus far, we have been focusing on "set-valued" choice functions. However, as noted in [27], a significant amount of the work in rational choice

---

[8]Sen discusses α2 and γ2 in connection with the "basic binary" choice functions. Given a choice function C on $\mathcal{X}$, let $\bar{R}_C$ be the binary relation on X that is defined as follows for all $x, y \in X$: $x R_C y$ iff $x \in C(\{x,y\})$. C is *basic binary* just in case $C(Y) = \{y \in Y \mid y\bar{R}_C z$ for all $z \in Y\}$ for all $Y \in \mathcal{X}$. As discussed in [28], C is basic binary iff it satisfies α2 and γ2.

has focused on "element-valued" choice functions. A *selection function* on X is a function $S : \mathcal{X} \to X$ such that $S(Y) \in Y$ for all $Y \in \mathcal{X}$.[9] It is clear that the selection functions on X may be regarded as a subset of the choice functions on X, since every such selection function S can be identified with the choice function C_S that is defined by $C_S(Y) = \{S(Y)\}$ for all $Y \in \mathcal{X}$. In light of this identification, conditions on choice functions may be interpreted as conditions on selection functions. In the case of selection functions it is known that α implies most of the familiar conditions that have been formulated for choice functions [27]. Using the suggested identification, α reduces to the following condition for selection functions: A selection function S on X satisfies α just in $S(Z) = S(Y)$ whenever $Y, Z \in \mathcal{X}$ and $S(Z) \in Y \subseteq Z$. It is well-known that α characterizes those selection functions that are generated by a linear order.[10]

Proposition 5.1. If S is a selection function on X, then S satisfies α iff there is a linear order $\succ$ on X such that, for all $Y \in \mathcal{X}$, there is no $y \in Y$ for which $y \succ S(Y)$.

Let us return to set-valued choice functions. We would like to use selection functions to characterize those choice functions within the suggested generalization that employ a vacuous security rule.[11] That is, given X, our target class consists of those choice functions C on X for which there exists a set $\mathcal{O}$ of orderings on X that generate C in the following sense: for all $Y \in \mathcal{X}$, $x \in C(Y)$ iff $x \in Y$ and there is an ordering $\succ$ in $\mathcal{O}$ such that there is no $y \in Y$ for which $y \succ x$.[12] We will now show that these functions are precisely the ones that are determined by the α-satisfying selection functions that they contain.

If S is a selection function on X and C is a choice function on X, then we will say that *S is contained in C* just in case, for all $Y \in \mathcal{X}$, $S(Y) \in C(Y)$. If $\mathcal{S}$ is a set of selection functions on X, then $\mathcal{S}$ *determines* C iff every element of $\mathcal{S}$ is contained in C and, for all $Y \in \mathcal{X}$, $C(Y) = \{S(Y) \mid S \in \mathcal{S}\}$. The following two propositions are immediate, but worth noting.

Proposition 5.2. Every choice function on X is determined by some set of selection functions on X.

---

[9] As before we let $\mathcal{X}$ denote the set of all finite, nonempty subsets of X.

[10] Recall that $P \subseteq X \times X$ is a linear order on X iff it satisfies the following conditions for all $x, y, z \in X$: (1) it is not the case that $x \succ x$, (2) if $x \neq y$, then either $x \succ y$ or $y \succ x$, and (3) if $x \succ y$ and $y \succ z$, then $x \succ z$.

[11] Of course, we would like to extend this approach by allowing for non-vacuous security rules, but here we restrict our attention to what looks to be a simpler problem and leave the more general case for future work.

[12] It is easy to see that every linear order is an order, although the converse does not hold since an order may have distinct elements that are not comparable.

Proposition 5.3. Let $\mathcal{S}$ be a set of selection functions on X. Let C and C' be choice functions on X. If $\mathcal{S}$ determines C and C', then $C = C'$.

Proposition 5.4. Let C be a choice function on X. If C is generated by an ordering, then C is determined by the set of α-satisfying selection functions that are contained in C.

Proof. Suppose that C is a choice function on X. Let $\succ$ be the ordering on X that generates C. Let $\sim$ be the binary relation on X defined by $x \sim y$ iff neither $x \succ y$ nor $y \succ x$. Recall that $\sim$ is an equivalence relation on X, since X is an ordering. Let P be a linear ordering of X.[13] Suppose that $x_0 \in X$ and $Y_0 \in \mathcal{X}$ are such that $x_0 \in C(Y_0)$. Let y_0 be the largest element in $C(Y_0)$ according to P. Note that such an element must exist since $C(Y_0)$ is finite. Let f be the function on X that is defined as follows: $f(x_0) = y_0$, $f(y_0) = x_0$, and $f(x) = x$ if $x \notin \{x_0, y_0\}$. Let P_f be the binary relation on X defined as follows for all $x, y \in X$: xP_fy iff $f(x)Pf(y)$. Note that P_f is a linear order on X since P is, and observe that x_0 is the largest element of $C(Y_0)$ according to P_f.

Define a binary relation $\succ^*$ on X as follows for all $x, y \in X$: $x \succ^* y$ iff $x \succ y$ or $x \sim y$ and xP_fy. We claim that $\succ^*$ is a linear order on X. Suppose that $x \succ^* x$. Since $\succ$ is an ordering, it cannot be the case that $x \succ x$. Hence, it must be the case that $x \sim x$ and xP_fx, but this is contradicts the fact that P_f is a linear order. Suppose that $x \succ^* y$ and $y \succ^* z$. We must show that $x \succ^* z$. If $x \succ y$ and $y \succ z$, then the result follows immediately from the transitivity of $\succ$. Suppose that $x \not\succ y$. Hence, $x \sim y$ and xP_fy. There are two subcases to consider. Assume that $y \succ z$. Since $x \sim y$ it follows that $y \not\succ x$. If $x \not\succ z$, then, by negative transitivity, $y \not\succ z$, which contradicts our assumption. Thus, $x \succ z$, and it follows that $x \succ^* z$. Assume that $y \sim z$ and yP_fz. Since $\sim$ and P_f are transitive, it follows that $x \sim z$ and xP_fz, and so $x \succ^* z$. The remaining cases are similar. Suppose that $x \neq y$. Observe that one of three cases obtains: $x \succ y$ or $y \succ x$ or $x \sim y$. If $x \succ y$, then $x \succ^* y$. If $y \succ x$, then $y \succ^* x$. If $x \sim y$, then either xP_fy, in which case $x \succ^* y$, or yP_fx, in which case $y \succ^* x$. Hence, if $x \neq y$, then either $x \succ^* y$ or $y \succ^* x$. This completes the argument to show that $\succ^*$ is a linear order on X.

Let S be the selection function that is generated by $\succ^*$. Observe that S is contained in C: Suppose that $Y \in \mathcal{X}$. It is enough to show that there is no $y \in Y$ such that $y \succ S(Y)$. Suppose that there is such a y. It follows that $y \succ^* S(Y)$, but this impossible since $S(Y)$ is the largest element of $S(Y)$ according to $\succ^*$. By construction, $S(Y_0) = x_0$. Also, by

---

[13]Recall that P is a well-ordering on X just in case it is a linear ordering on X such that every nonempty subset of X has a least element with respect it. We assume the axiom of choice. Hence, there is well-ordering of X.

Proposition 5.1, S satisfies α. Since nothing was assumed about x_0 and Y_0, other than $x_0 \in C(Y_0)$, there is, for each pair (x, Y) such that $x \in C(Y)$, an α-satisfying selection function that is contained in C and selects x from Y. It follows that C is determined by the set of α-satisfying selection functions that it contains. Q.E.D.

Proposition 5.5. Let C be a choice function on X. If C is generated by a set of orderings, then C is determined by the set of α-satisfying selection functions that are contained in C.

Proof. Suppose that C is generated by a set of orderings. It follows that $C = \bigcup_{i \in I} C_i$ for some set of choice functions $\{C_i\}_{i \in I}$ where each C_i is generated by an ordering. By Proposition 5.4, each C_i is determined by the set of α-satisfying selection functions that it contains. It is easy to see that C is determined by the union of these sets and that this union is a subset of the set of all α-satisfying selection functions that are contained in C. The desired conclusion follows immediately. Q.E.D.

Proposition 5.6. Let C be a choice function on X. If C is is determined by the set of α-satisfying selection functions that are contained in C, then C is generated by a set of orderings.

Proof. Suppose that C is determined by the set $\mathcal{S}$ of all α-satisfying selection functions that are contained in C. By Proposition 5.1, every $S \in \mathcal{S}$ can be identified with a linear ordering $\succ_S$ on X. It is easy to see that $\mathcal{O} = \{\succ_S | S \in \mathcal{S}\}$ is a set of orderings that generates C. Q.E.D.

Proposition 5.7. Let C be a choice function on X. If C is determined by the set of α-satisfying selection functions that are contained in C, then C satisfies α.

Proof. Suppose that C is determined by the set $\mathcal{S}$ of all α-satisfying selection functions that are contained in C. Suppose that $Y, Z \in \mathcal{X}$ where $x \in Y \subseteq Z$. If $x \in C(Z)$, then, since C is determined by $\mathcal{S}$, there is an $S \in \mathcal{S}$ such that $S(Z) = x$. Since S satisfies α, it follows that $S(Y) = x$. Since S is contained in C, it follows that $x \in C(Y)$. Hence, C satisfies α. Q.E.D.

The converse of Proposition 5.7 is false. Consider the choice function C on $\{a, b, c\}$ defined as follows: $C(\{a, b, c\}) = \{a\}$, $C(\{a, b\}) = \{a\}$, $C(\{b, c\}) = \{b\}$, and $C(\{a, c\}) = \{a, c\}$. C satisfies α. However, C is not determined by the set of α-satisfying selection functions that it contains: C contains two selection functions, resulting from the two possible selections at $\{a, c\}$. Both of these selection functions are needed to determine C. However, the one that selects c from $\{a, c\}$ does not satisfy α.

References

[1] Mark A. Aizerman. New Problems in the General Choice Theory. *Social Choice and Welfare*, 2:235–282, 1985.

[2] Mark A. Aizerman and Andrei V. Malishevski. General Theory of Best Variants Choice: Some Aspects. *IEEE Transactions on Automatic Control*, 26:1030–1040, 1981.

[3] Daniel Ellsberg. Risk, Ambiguity, and the Savage Axioms. *Quarterly Journal of Economics*, 75:643–669, 1961.

[4] Daniel Ellsberg. *Risk, Ambiguity and Decision*. Garland Publishing, 2001.

[5] Peter C. Fishburn. Subset Choice Conditions and the Computation of Social Choice Sets. *Quarterly Journal of Economics*, 88:320–329, 1974.

[6] Craig R. Fox and Amos Tversky. Ambiguity Aversion and Comparative Ignorance. *Quarterly Journal of Economics*, 110:585–603, 1991.

[7] Peter Gardenfors and Nils E. Sahlin. Unreliable Probabilities, Risk Taking, and Decision Making. *Synthese*, 53:361–386, 1982.

[8] Itzhak Gilboa and David Schmeidler. Maximin Expected Utility with Non-unique Prior. *Journal of Mathematical Economics*, 18:141–153, 1989.

[9] Jeffrey Helzner. On the Application of Multiattribute Utility Theory to Models of Choice. *Theory and Decision*, 66:301–315, 2009.

[10] Richard C. Jeffrey. Decision Theory. In Robert Audi, editor, *Cambridge Dictionary of Philosophy*, pages 179–181. Cambridge University Press, 1995.

[11] Barbara E. Kahn and Robert J. Meyer. Consumer Multiattribute Judgments under Attribute-Weight Uncertainty. *Journal of Consumer Research*, 17:508–522, 1991.

[12] Ralph L. Keeney and Howard Raiffa. *Decisions with Multiple Objectives: Preferences and Value Tradeoffs*. Cambridge University Press, 1993. The Cambridge edition is a republication of the 1976 work.

[13] John Maynard Keynes. *A Treatise on Probability*. MacMillan, 1921.

[14] Frank H. Knight. *Risk, Uncertainty and Profit*. Houghton-Mifflin, 1921.

[15] David Krantz, Duncan Luce, Patrick Suppes, and Amos Tversky. *Foundations of Measurement Volume I: Additive and Polynomial Representations*. Dover Publications, 1971.

[16] David Kreps. *Notes on the Theory of Choice*. Westview, 1988.

[17] Isaac Levi. On Indeterminate Probabilities. *Journal of Philosophy*, 71:391–418, 1974.

[18] Isaac Levi. *Hard Choices: Decision Making under Unresolved Conflict*. Cambridge University Press, 1986.

[19] Isaac Levi. The Paradoxes of Allais and Ellsberg. *Economics and Philosophy*, 2:23–53, 1986.

[20] Duncan Luce and Howard Raiffa. *Games and Decisions: Introduction and Critical Survey*. Dover Publications, 1989. The Dover edition is a republication of the 1957 work.

[21] David Mellor, editor. *Frank P. Ramsey. Philosophical Papers*. Cambridge University Press, 1990.

[22] Klaus Nehring. Rational Choice and Revealed Preference without Binariness. *Social Choice and Welfare*, 14:403–425, 1997.

[23] Klaus Nehring and Clemens Puppe. Extending Partial Orders: A Unifying Structure for Abstract Choice Theory. *Annals of Operations Research*, 80:27–48, 1998.

[24] Frank P. Ramsey. Truth and Probability. In Richard B. Braithwaite, editor, *Frank P. Ramsey, The Foundations of Mathematics and other Logical Essays*, pages 156–198. Kegan, Paul, Trench, Trubner & Co., 1931. Reprinted in [21, pp. 52–109].

[25] Leonard J. Savage. *The Foundations of Statistics*. Dover Publications, 1972. The Dover edition is a republication of the 1954 work.

[26] Teddy Seidenfeld, Mark J. Schervish, and Joseph B. Kadane. Coherent Choice Functions under Uncertainty. In Gert de Cooman, Jiřina Vejnarová, and Marco Zaffalon, editors, *Proceedings of the Fifth International Symposium on Imprecise Probability: Theories and Applications*. Action M. Agency for SIPTA, 2007.

[27] Amartya Sen. Choice Functions and Revealed Preference. *Review of Economic Studies*, 38:307–317, 1971.

[28] Amartya Sen. Social Choice Theory: A Re-Examination. *Economet-rica*, 45:53–88, 1977.

[29] A. Tversky and D. Kahneman. Judgment under Uncertainty: Heuristics and Biases. *Science*, 185:1124–1131, 1974.

[30] Peter P. Wakker. *Additive Representation of Preferences*. Kluwer Academic Publishers, 1989.

[31] Martin Weber. A Method of Multiattribute Decision Making with Incomplete Information. *Management Science*, 31:1365–1371, 1985.

Benedikt **Löwe**, Eric **Pacuit**, Jan-Willem **Romeijn** (*eds.*)
Foundations of the Formal Sciences VI
Reasoning about Probabilities and Probabilistic Reasoning

Logical Relations in a Statistical Problem

ROLF HAENNI[1,2], JAN-WILLEM ROMEIJN[3],
GREGORY WHEELER[4], JON WILLIAMSON[5]

[1] Departement Technik und Informatik, Berner Fachhochschule, Höheweg 80, 2501 Biel, Switzerland

[2] Institut für Informatik und angewandte Mathematik, Universität Bern, Neubrückstrasse 10, 3012 Bern, Switzerland

[3] Faculteit Wijsbegeerte, Rijksuniversiteit Groningen, Oude Boteringestraat 52, 9712 GL Groningen, The Netherlands

[4] Centre for Research in Artificial Intelligence, Departamento de Informática, Faculdade de Ciências e Tecnologia, Universidade Nova de Lisboa, Quinta da Torre, 2829-516 Caparica, Portugal

[5] Philosophy Section, School of European Culture & Languages, The University of Kent, Canterbury, Kent, CT2 7NZ, United Kingdom

E-mail: `rolf.haenni@bfh.ch`, `J.W.Romeijn@rug.nl`, `grw@fct.unl.pt`, `j.williamson@kent.ac.uk`

1   Introduction

While in principle probabilistic logics might be applied to solve a range of problems, in current practice they are rarely applied. This is perhaps because they seem disparate, complicated, and computationally intractable. In fact, as we shall illustrate in this paper, several approaches to probabilistic logic fit into a simple unifying framework. Furthermore, there is the potential to develop computationally feasible methods to mesh with this framework. A unified framework for dealing with logical relations may contribute to probabilistic methods in machine learning and statistics, much in the way that the notion of causality and its relation to Bayesian networks have contributed to advances in these fields.

The unifying framework is developed in detail in [6]. Here we shall very briefly describe the gist of the whole approach.

Received by the editors: 8 October 2007; 1 May 2008.
Accepted for publication: 5 May 2008.

1.1   Probabilistic Logic

Probabilistic logic asks what probability (or set of probabilities) should attach to a conclusion sentence ψ, given premises which assert that certain probabilities (or sets of probabilities) attach to various sentences $\varphi_1, \ldots, \varphi_n$. That is, the fundamental question is to find a suitable set Y such that

$$\varphi_1^{X_1}, \ldots, \varphi_n^{X_n} \approx \psi^Y, \tag{1.1}$$

where $\approx$ is a notion of entailment, $X_1, \ldots, X_n, Y$ are sets of probabilities and $\varphi_1, \ldots, \varphi_n, \psi$ are sentences of some logical language $\mathcal{L}$. This is a *schematic* representation of probabilistic logic, inasmuch as the entailment relation $\approx$ and the logical language $\mathcal{L}$ are left entirely open.

1.2   The Progicnet Programme

What we call the *progicnet programme* consists of two basic claims:

Framework. A unifying framework for probabilistic logic can be constructed around Schema 1.1;

Calculus. Probabilistic networks can provide a calculus for probabilistic logic—in particular they can be used to find a suitable Y such that the entailment relation of Schema (1.1) holds.

These two claims offer a means of unifying various approaches to combining probability and logic in a way that seems promising for practical applications. We shall now take a look at these two claims in more detail.

1.2.1   Framework

The first claim is that a unifying framework for probabilistic logic can be constructed around Schema (1.1). This claim rests on the observation that several seemingly disparate approaches to inference under uncertainty can in fact be construed as providing semantics for Schema (1.1):

Standard Probabilistic Semantics. According to the standard semantics, the entailment $\varphi_1^{X_1}, \ldots, \varphi_n^{X_n} \approx \psi^Y$ holds if all probability functions P which satisfy the premises—i.e., for which $P(\varphi_1) \in X_1$, $\ldots$, $P(\varphi_n) \in X_n$—also satisfy the conclusion $P(\psi) \in Y$. The logical language may be a propositional or predicate language.

Bayesian Statistical Inference. Under this account, the probabilistic premises contain information about prior probabilities and likelihoods which constitute a statistical model, the conclusion denotes posterior probabilities, and the entailment holds if, for every probability function subsumed by the statistical model of the premises, the conclusion follows by Bayes's theorem. Again a propositional or predicate language may be used.

Evidential Probability. Here the language is a predicate language that can represent statistical statements of the form 'the frequency of S in reference class R is between l and u'. The φ_i capture the available evidence, which may include statistical statements. These evidential statements are uncertain and the X_i characterise their associated risk levels. The entailment holds if the conclusion follows from the premisses by the axioms of probability and certain rules for manipulating statistical statements.

Probabilistic Argumentation. Here the language is propositional and the entailment holds if Y contains the proportion of worlds for which the left-hand side forces ψ to be true.

Objective Bayesian Epistemology. This approach deals with a propositional or predicate language. The $\rho_i^{X_i}$ are interpreted as evidential statements about empirical probability, and the entailment holds if the most *non-committal* (i.e., maximum entropy) probability function, from all those that that satisfy the premisses, satisfies the conclusion.

With the exception of the first, these different semantics for probabilistic logic are presented more fully in the subsequent sections of this paper.

1.2.2   Calculus

In order to answer the fundamental question that a probabilistic logic faces—i.e., in order to find a suitable Y—some computational machinery needs to be invoked. Rather than appealing to a proof theory as is usual in logic, the progicnet programme appeals to *probabilistic networks*. This is because determining Y is essentially a question of probabilistic inference, and probabilistic networks can offer a computationally tractable way of inferring probabilities. It turns out that under the different approaches to probabilistic inference outlined above, it is often the case that $X_1, \ldots, X_n, Y$ are single probabilities or intervals of probability. When that is the case, a *Bayesian network* (a tool for drawing inferences from a single probability function) or a *credal network* (which draws inferences from a closed convex set of probability functions) can be used to determine Y. The construction of the probabilistic network depends on the chosen semantics, but given the network the determination of Y is independent of semantics. Hence the progicnet programme includes a common set of tools for calculating Y [6]. Examples of the use of probabilistic networks will appear in the following sections; here we shall introduce the key features of probabilistic networks and their role in the progicnet programme.

A probabilistic network is based around a set of variables $\{A_1, \ldots, A_r\}$. In the context of probabilistic logic, these may be propositional variables, taking two possible values True or False; if the language $\mathcal{L}$ of the logic is a predicate language, the propositional variables may represent atomic propositions, i.e., propositions of the form Ut where U is a relation symbol and t is a tuple of constant symbols. A probabilistic network contains a directed acyclic graph whose nodes are $A_1, \ldots, A_r$. This graph is assumed to satisfy the *Markov condition*: each variable is probabilistically independent of its non-descendants, conditional on its parents in the graph. For instance, the following directed acyclic graph implies that A_3 is independent of A_1 conditional on A_2:

FIGURE 1. Example of a probabilistic network.

A probabilistic network also contains information about the probability distribution of each variable conditional on its parents in the graph. In a Bayesian network, these conditional probabilities are all fully specified; a Bayesian network then determines a joint probability distribution over $A_1, \ldots, A_r$ via the relation $\mathrm{P}(A_1, \ldots, A_r) = \prod_{i=1}^{r} \mathrm{P}(A_i|\mathrm{Par}_i)$ where Par_i is the set of parents of A_i. In our example, we might have

$$\mathrm{P}(A) = 0.7, \qquad \mathrm{P}(B|A) = 0.2, \qquad \mathrm{P}(C|B) = 0.9,$$
$$\mathrm{P}(B|\neg A) = 0.1, \qquad \mathrm{P}(C|\neg B) = 0.4,$$

from which we derive, for example,

$$\mathrm{P}(A \wedge \neg B \wedge C) = \mathrm{P}(A)\mathrm{P}(\neg B|A)\mathrm{P}(C|\neg B) = 0.224.$$

In a credal network, the conditional probabilities are only constrained to lie within closed intervals. A credal network then determines a set of joint probability distributions: the set of those distributions determined by Bayesian nets that satisfy the constraints. For example, a credal network might by satisfied by the above graph together with the following constraints:

$$\mathrm{P}(A) \in [0.7, 0.8], \qquad \mathrm{P}(B|A) = 0.2, \qquad \mathrm{P}(C|B) \in [0.9, 1],$$
$$\mathrm{P}(B|\neg A) \in [0.1, 1], \qquad \mathrm{P}(C|\neg B) \in [0.4, 0.45].$$

In the context of probabilistic logic, we are given premises $\varphi_1^{X_1}, \ldots, \varphi_n^{X_n}$, and a conclusion sentence ψ, and we need to determine an appropriate Y to attach to ψ. The idea is to build a probabilistic network that represents the set of probability functions satisfying the premises, and use this network to calculate the range of probabilities that these functions give ψ. As mentioned above, the construction of the probabilistic network will depend on the chosen semantics, but common inference machinery may be used to calculate Y from this network. The approach taken in [6, §8.2] is to implement this common machinery as follows. First, *compile* this network: i.e., transform it into a different kind of network which is guaranteed to generate inferences in an efficient way. Second, use numerical *hill-climbing* methods in this compiled network to generate an approximation to Y.

In this paper we will illustrate the general approach of the progicnet programme by means of an example in which a number of applications can be exhibited. The example stems from psychology, more specifically from psychometrics, which studies the measurement of psychological attributes by means of tests and statistical procedures performed on test statistics. This example is constructed with the aim of bringing out the use of logical relations in probabilistic inference. In the next section we shall introduce the psychometric case study. In subsequent sections we shall see how the inferential procedures introduced above can be applied to this problem domain, and how they fit into a single framework within which the progicnet calculus can be utilized.

2 Applying the Progicnet Framework

We now illustrate the progicnet programme with an example on the measurement of psychological attributes. The first subsection introduces the example, and the second subsection indicates how each of the approaches that is covered by the progicnet framework can be employed to solve specific problems. At times, the example may come across as somewhat contrived. If so, this is because we illustrate all procedures with a single example. Straightforward applications of the framework and calculus will typically involve two procedures only.

2.1 A Psychometric Case Study

Psychometrics is concerned with the measurement of psychological attributes in individuals, for example to do with cognitive abilities, emotional states, and social strategies. Typically, such attributes cannot be observed directly. What we observe are the behavioural consequences of certain psychological attributes, such as a high score in a memory test, a certain reaction to emotionally charged images, or the characteristics of social interactions in some game. In many psychometric studies, the psychological attributes are taken as the hidden causes of these observable facts about subjects, or in short, they are taken as *latent variables*. The *observable variables*, and the correlational structure among them, are used to derive facts about these latent variables.

Notice that the general aim of psychometrics fits well with the general outlook of the progicnet framework. As in the progicnet framework, most psychometric questions start out with a number of probabilistic facts, deriving directly from the observations, and a number of logical and probabilistic relations among observable and latent variables, deriving from the psychological theory. The goal is then to find further logical and probabilistic facts concerning the latent variables, which satisfy the constraints determined by the observations and the psychological theory. Hence psychometrics lends itself well to a conceptualisation in terms of the progicnet framework.

Let us make this more concrete in the context of a version of a cognitive psychological experiment, which we concede is still rather abstract. Say that we have presented a number of subjects, indexed j, with three cognitive ability tasks, A, B, and C, which they can either pass or fail. We denote the corresponding test variables by A_j, B_j, and C_j, denoting the scores of subjects j on the three tests, respectively. Each test variable can be true or false, which, in the case of A_j, is denoted by the assignments a_j^1 (or a_j) and a_j^0 (or $\neg a_j$), respectively.

Imagine further that these tests are supposed to inform us about a psychological theory concerning three aspects of cognition, two of them to do with different developmental stages of the subjects and the other with processing speed. The corresponding latent variables are denoted by F_j, G_j, and H_j, respectively. Say that the categorical variables F_j and G_j each discern two developmental stages, and are thus binary. The processing speed $H_j \in [0, \infty)$ is continuous, but for convenience we may view H_j as categorical on some suitable scale, taking integer values n for $1 \leq n \leq N$ and N sufficiently large, say $N = 100$. The atomic statements in the language are then valuations of these variables for subjects. For example, b_5^0 or $\neg b_5$ mean that subject $j = 5$ failed test B, and h_3^{15} means that subject $j = 3$ has a latent processing speed $n = 15$. For convenience we collect the variables in $V_j = \{A_j, B_j, C_j, F_j, G_j, H_j\}$.

Imagine first that the psychological theory provides the following independence relation among the variables in the theory:

$$\forall j \neq k : \quad \mathrm{P}(V_j) = \mathrm{P}(V_k). \tag{2.1}$$

This relation expresses that all subjects are on the same footing, in the sense that they are each described by the same probability function over all the variables. Because of this the order in which the subjects are sampled does not matter to the conclusions we can draw from the sample. Moreover, unless we condition on observations of specific subjects and assignments, we can omit reference to the subjects j in the probability assignments to the variables.

Second, imagine that the developmental stages F and G and processing speed H are independent components in determining the test performance, and further that test scores are determined only by these latent variables, i.e., conditional on the latent variables, the performance on the tests is uncorrelated. The exact independence structure might be:

$$P(A, B, C, F, G, H) = P(F)P(G)P(H)P(A|F, G)P(B|G, H)P(C|H). \quad (2.2)$$

Both the independence among the subjects j, and the independence relations between the variables within each subject present strong simplifications to the psychometric example.

Next to the independence premises, psychological theory might determine the following relations between assignments to the latent and the observable variables. All these relations hold for all subjects j, and thus we omit again any such reference.

$$f \wedge g \rightarrow \neg a, \quad (2.3)$$

$$\neg g \rightarrow a, \quad (2.4)$$

$$P(b|g \wedge h^n) = \frac{n}{N}, \quad (2.5)$$

$$P(c|h^n) = \frac{N + n}{2N}. \quad (2.6)$$

Again these relations may be taken as premises in the progicnet framework, because each of these relations effectively restricts the set of probability assignments over both latent and observable variables. Or in terms more familiar to statisticians, the above premises determine a model: they fix the likelihoods of the hypotheses about subjects. Note, however, the available knowledge about the outcome of test A, as expressed in Equations (2.3) and (2.4), is purely logical and in this sense qualitative. One of the challenges is to combine such purely logical constraints with the probabilistic facts given in the other premises.

2.2 Various Approaches in a Unifying Framework

As signalled at the beginning of this section, the reader may feel that the psychometric example is unnecessarily complicated. We hope it will be apparent from subsequent sections why the example is so multi-faceted. One of the strengths of the progicnet framework is that it can accommodate a large variety of inferential problems, and we have chosen the example such that all these inferential problems find a natural place.

Of course a large number of problems on the psychometric example are essentially statistical. We may want to estimate the probability that a subject will pass test C given her performance on A and B, or how probable it is that her processing speed exceeds a certain threshold. Most of these problems will be dealt with in Bayesian statistical inference, which is sketched in Section 3. There we define a probability over the latent variables, by observing a number of subjects and then adapting the probability over latent variables accordingly. Because this type of inference is particularly well-suited for the example, we will pay a fair amount of attention to it.

Of course there are also statistical inference problems to which Bayesian statistical inference is not that easily applicable. For example, we might discover that an additional factor D influences the performance on the tests A, B, and C, so that we have to revise our predictions over these performances. Alternatively, we might be given further frequency information from various experimental studies on the variables already present in the example, say

$$P(g|b) \in [0.2, 0.4], \tag{2.7}$$

$$P(g|c) \in [0.3, 0.5]. \tag{2.8}$$

On the addition of such information, we can employ inferences that use so-called evidential probability. It tells us how to employ the discovery of the factor D in improving predictions, and how to adapt the predictions for G after learning the further frequency information. Section 4 introduces this approach.

Evidential probability provides solutions to a number of inferential problems on which Bayesian inference remains silent. But there are yet other problems for which both these statistical approaches are unsuited, for instance those concerned with logically complex combinations of observable and latent variables. Say that growing theoretical insight entails that

$$(a \wedge g) \vee b. \tag{2.9}$$

We might then ask what probability to attach to other complex formulae. As worked out in Section 5, probabilistic argumentation is able to provide answers on the basis of a strict distinction between logical and probabilistic knowledge, and by considering the probability of a hypothesis to be deducible from the given logical premises. However, answers to such questions will typically be intervals of probability, which makes actual computations less efficient. Here objective Bayesianism, as dealt with in Section 6, presents a technique to select a single probability assignment from all assignments that are consistent with the premises.

In the next few sections we show that inferential problems such as the above can be answered by the variety of approaches alluded to in the above, that these approaches can all be accommodated by the progicnet framework, and that their accommodation by the framework makes them amenable to

the common calculus introduced in the foregoing. In this way we illustrate
the use of this framework.

3   Bayesian Statistical Inference

This section introduces Bayesian statistical inference, illustrates how it is
captured in the progicnet framework, and finally shows that it can be employed to solve inferential problems on the psychometric example. Bayesian
statistical inference is a relatively important approach in this paper. It covers a fairly large number of the inferential problems in the example, because
the example itself has a statistical nature. However, it also misses important aspects. In subsequent sections, we will show how each of the other
approaches in this paper can be used to fill in these lacunas.

3.1   Simple Bayesian Inference in the Progicnet Framework

The key characteristic of Bayesian statistics is that it employs probability assignments over statistical hypotheses, next to probability assignments
over data. More specifically, a Bayesian statistical inference starts by determining a model, or a *set of statistical hypotheses* that are each associated
with a full probability assignment over the data, otherwise known as *likelihood functions*, and further a so-called *prior probability assignment* over the
model. Relative to a model and a prior probability, the data then determine
a so-called *posterior distribution* over the model, and from this posterior we
can derive expectation values, predictions, credence intervals, and the like
[1, 13].

We may illustrate the general idea of Bayesian inference with the psychometric example of the previous section. In the example, $\{h_j^1, \ldots, h_j^N\}$ is
a model with a finite number of hypotheses concerning the latent speed of
some subject j, and Equation (2.6) determines the likelihoods $P(c_j^1 | h_j^n) =
\frac{N+n}{2N}$ of the hypotheses h_j^n for c_j^1, the event of subject j passing the test
C_j. Finally, we might take a uniform distribution $P(h_j^n) = \frac{1}{N}$ as prior
probabilities. With Bayes's theorem it follows that

$$P(h_j^n | c_j^1) = P(h_j^n) \frac{P(c_j^1 | h_j^n)}{P(c_j^1)} = \frac{2(N+n)}{N(3N+1)}. \tag{3.1}$$

That is, upon learning that subject j passed test C_j, we may adapt the
probability assignment over processing speeds for that subject to the values
on the right hand side. This transition from the prior $P(h_j^n)$ to the posterior
$P(h_j^n | c_j^1)$ is at the heart of all Bayesian statistical inferences.

It may be noted that the probability of the datum $P(c_j^1)$ appears in
Bayes's theorem. This probability may seem hard to determine directly.

However, by the law of total probability we have

$$P(c_j^1) = \sum_{n=1}^{N} P(h_j^n)P(c_j^1|h_j^n) = \frac{3N+1}{4N}. \tag{3.2}$$

So, relative to a model, the probability of c_j^1 is easily determined. We simply need to weigh the likelihoods of the hypotheses with the prior over the hypotheses in the model.

We can represent the transition from prior and likelihoods to posterior in the progicnet framework, as it was introduced in Section 1. Recall that in Schema (1.1), all premises take the form of restrictions to a probability of a logical expression, $\varphi_i^{X_i}$. However, the likelihoods $P(c_j^1|h_j^n) = \frac{N+n}{2N}$ cannot be identified directly with probability assignments to specific statements, because $c_j^1|h_j^n$ does not correspond to a specific proposition. They do represent restrictions to the probability assignments, but rather they are restrictions of a different type. Since

$$P(c_j^1|h_j^n) = \frac{P(c_j^1 \wedge h_j^n)}{P(h_j^n)},$$

we can define this restriction as the combination of two related and direct restrictions to the probability assignments, as follows:

$$\left(c_j^1|h_j^n\right)^{\frac{N+n}{2N}} \Leftrightarrow \forall \gamma \in [0,1] : \left(h_j^n\right)^\gamma \text{ and } \left(c_j^1 \wedge h_j^n\right)^{\gamma \frac{N+n}{2N}}. \tag{3.3}$$

The left side of this equivalence is the likelihood in the notation of Schema (1.1), while the right side fixes the probability of two related propositions in parallel. In words, we restrict the set of probability functions over the algebra to those functions for which the ratio of the probabilities of the two propositions $c_j^1 \wedge h_j^n$ and h_j^n is $\frac{N+n}{2N}$.

With this notation in place, all expressions in Equation (3.1) are seen to be restrictions to a class of probability assignments, or models for short. More specifically, the restrictions together determine the set of models uniquely: only one probability assignment over the h_j^n's and c_j^1's satisfies the restrictions on the left hand side. But this is not to say that the complete credal set, as introduced in Section 1, is a singleton. The one probability assignment over the h_j^n's and c_j^1's can still be combined with any probability assignment over the other propositional variables.

Still restricting attention to the transition from prior to posterior for the hypotheses h_j^n and the data c_j^1, the Bayesian inference can now be represented straightforwardly in the form of Schema (1.1):

$$\forall n \in \{1, \ldots, N\} : (h_j^n)^{\frac{1}{N}}, (c_j^1|h_j^n)^{\frac{N+n}{2N}} \models (h_j^n|c_j^1)^{\frac{2(N+n)}{N(3N+1)}}. \tag{3.4}$$

Equation (3.4) is a representation of the Bayesian statistical inference, starting with a model of hypotheses h_j^n, their priors and likelihoods

$$P(h_j^n) = \frac{1}{N}, \qquad P(c_j^1|h_j^n) = \frac{N+n}{2N},$$

and ending with a posterior

$$P(h_j^n|c_j^1) = \frac{2(N+n)}{N(3N+1)}.$$

The derivation of the posterior employs standard probability theory and concerns credal sets. It is therefore amenable to the calculus introduced in Section 1.

In sum, provided we supply the relevant premises, we can also interpret inferences within the progicnet framework as Bayesian statistical inferences. One type of premise concerns the statistical model, the other type of premise determines the prior probability assignment over the model. From these two sets of restrictions we can derive, by using the progicnet calculus, a further restriction on the posterior probability $P(h_j^n|c_j^1)$.

3.2   Bayesian Inference across Subjects

The above makes explicit what Bayesian statistical inference is, and how it relates to the progicnet framework. In the remainder of this section, we will show that we can accommodate the psychometric example in its entirety in a Bayesian statistical inference. That is, we extend Bayesian inference to apply to all variables and subjects, and we include all probabilistic restrictions presented in the example. It is noteworthy that this involves additional assumptions to do with a prior over latent and observable variables. If we want to do without such assumptions, we must move to one of the other approaches for incorporating logical and probabilistic relations that this paper deals with.

Recall that the idea of statistical inference is not just that we can learn about values of variables *within subjects*, but that we can learn about them *across subjects*. For example, from observing the value of C_j for a subject j we should be able to derive something about the probability assignment over the values H_k for a different subject k. The independence expressed in Equation (2.2) determines in what way this learning across subjects can take place. It expresses that each subject has a valuation over both latent and observable variables, that is drawn from the same multinomial distribution $P(V)$ with $V = \{A, B, C, F, G, H\}$. By learning valuations and expectations over these variables for some subjects, we therefore also learn the expectations over variables for other, as yet unobserved subjects. Moreover, the valuations of the variables are not drawn from just any multinomial distribution over the variables. Because we only have access to the observable variables, the latter would mean we could never learn anything about the latent variables. Fortunately the psychometric example offers a number of relations among latent and observable variables, and these relations restrict the set of multinomial distributions from which the valuations of the observable variables are drawn.

To make this specific, consider again the relation between the observable variables C_j and the latent variables H_j. To keep things manageable we choose $N = 3$, so that we have $3 \times 2 = 6$ complete valuations of C_j and H_j together. Without further restrictions, we thus have a multinomial distribution determined by 6 parameters, namely probabilities for each full valuation $P(c_j^i \wedge h_j^n) = \theta_k$ with $k = ni + n$, and a restriction that these probabilities add up to 1, leading to 5 degrees of freedom. We can also parameterise this distribution differently, with a probability $P(h_j^n) = \theta_{h^n}$ for the latent variables h^n, a restriction that these sum to 1, and next to that three conditional probabilities $P(c_j^1|h_j^n) = 1 - P(c_j^0|h_j^n) = \theta_{C^n}$. In either case we have a set of multinomial distributions from which valuations of observed and latent variables may be drawn.

As suggested in the foregoing, we have some additional restrictions to this set of distributions deriving from the likelihoods of H_j for C_j: $P(c_j^1|h_j^n) = \frac{N+n}{2N}$. In the latter parametrisation of the multinomial distributions, these restrictions can be accommodated very easily, because they come down to setting parameters θ_{C^n} to specific values, namely

$$\theta_{C^n} = \frac{N+n}{2N}. \tag{3.5}$$

Once the restrictions given by the likelihoods $P(c_j^1|h_j^n)$ are put in place, all remaining degrees of freedom in the parameter space derive from the freedom in the probability over the hypotheses $P(h_j^n)$. Every point in the parameter space $\theta_h = \langle \theta_{h^1}, \theta_{h^2}, \theta_{h^3} \rangle$ is associated with a particular value for the probability of the observable variable C_j, according to

$$P_{\theta_h}(c_j^1) = \sum_{n=1}^{3} P(c_j^1|h_j^n)P(h_j^n) = \sum_{n=1}^{3} \frac{N+n}{2N}\theta_{h^n}. \tag{3.6}$$

Note that these values need not be unique: it may happen, and indeed it does happen in the example, that several probability assignments over the h_j^n, or points θ_h in the parameter space, lead to the same overall probability for c_j^1. Hence observing the relative frequency of values for the variables C_j may not lead to a unique probability over the hypotheses $P(h_j^n)$. In any case, the main insight is that learning the relative frequency of values for the variables C_j does tell us something about the probabilities of h_k^n for some as yet unobserved subject k.

3.3 Setting up the Statistical Model

The foregoing concludes the introduction into Bayesian statistical inference for the psychometric example. We will now fill in the details of this approach. The aim is to specify a Bayesian inference for F, G and H from the observation of A, B, and C and the relations (2.3) to (2.6), along the lines just sketched for H and C. Readers who are more interested in the complementary tools provided by the other approaches can skip the present subsection.

As indicated in Section 1, to make actual inferences in the psychometric example it is convenient to build up a so-called *credal network*, a graphical representation of the probability assignment over all the variables, and to build up the parametrisation of the multinomial distribution, from which observations are drawn, on the basis of this network. By the independence relation of Equation (2.2) we have the following network:

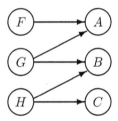

FIGURE 2. The network for the psychometric case study.

This network captures the independence relations for each subject j separately. It expresses exactly the independencies brought out by Equation (2.2): conditional on certain latent variables certain test variables are independent of each other, and the three latent variables are independent of each other as well.

Now that we have pinned down this overall structure of the model, we can fill in some of the details by means of the relations between latent and observable variables. More specifically, from Equation (2.4) we can derive that

$$g_j^0 \wedge a_j^0$$

is false, so that we have $P(a_j^0|g_j^0) = 0$ and hence

$$P(a_j^0|g_j^0 \wedge f_j^i) = 0$$

for $i = 0, 1$. Similarly, from Equation (2.3) we can derive that $f_j^1 \wedge g_j^1 \wedge a_j^1$ is false, so that we have

$$P(a_j^1|g_j^1 \wedge f_j^1) = 0.$$

Equations (2.5) and (2.6) provide input to the Bayesian inference even more straightforwardly: they fix the values for $P(b_j^1|g_j^1 \wedge h_j^n)$ and $P(c_j^1|h_j^n)$ respectively. The nice thing about the above network representation is that its parametrisation, in terms of probabilities for latent variables and probabilities of observable variables conditional on these latent variables, allows us to include these restrictions directly. All the relations between latent and observable variables restrict the space of multinomial probability distributions, by setting one or more of its parameters to specific values.

After all these relations have been incorporated, we have narrowed down the set of multinomial distributions to a specific set, which we may denote $\mathbb{P}$. Within this specific set, we have the following degrees of freedom left:

$$P(a_j^1|f_j^0 \wedge g_j^1) = \theta_{A^1|F^0G^1}, \tag{3.7}$$

$$P(b_j^1|g_j^0 \wedge h_j^n) = \theta_{B^1|G^0H^n}, \tag{3.8}$$

$$P(f_j^1) = \theta_{F^1}, \tag{3.9}$$

$$P(g_j^1) = \theta_{G^1}, \tag{3.10}$$

$$P(h_j^n) = \theta_{h^n}. \tag{3.11}$$

So for $N = 3$ we have 7 degrees of freedom left in the space of multinomial distributions. Note that the uncertainty of the likelihoods, Equations (3.7) and (3.8), is quite different from the uncertainty over the latent variables, Equations (3.9) to (3.11). The former uncertainty concerns the evidential bearing that the observable variables have on the latent variables, while the latter uncertainties concern the latent variables themselves.

For each point within the above space of multinomial distributions, we can derive likelihoods for the observable variables A and B, analogously to Equation (3.6) for C:

$$P(a_j^1) = (1 - \theta_{F^1})\,\theta_{G^1}\,\theta_{A^1|F^0G^1}, \tag{3.12}$$

$$P(b_j^1) = \sum_{n=1}^{3} \theta_{h^n} \left(\theta_{G^1}\frac{n}{N} + (1 - \theta_{G^1})\theta_{B^1|G^0H^n}\right). \tag{3.13}$$

Because Equations (2.3) to (2.6) do not pin down all evidential relations, the likelihoods for A_j and B_j will also depend on the values of $\theta_{A^1|F^0G^1}$ and $\theta_{B^1|G^0H^n}$. One possible reaction to this is that we stipulate specific values for the latter parameters, for instance by the maximum entropy principle. This approach is developed further in Section 6.

The fully Bayesian reaction, however, is to include the unknown likelihoods in the space of multinomial distributions, and to work with a second-order probability assignment over the entire space, which includes parameters pertaining to the probability of latent variables, and parameters pertaining to observable variables conditional on latent variables. We then assign a prior probability assignment to each point in the space of multinomial distributions. And once we have provided a prior probability over all parameters, we can integrate the parameters $\theta_{A^1|F^0G^1}$ and $\theta_{B^1|G^0H^n}$ out, and come up with a marginal likelihood for A_j and B_j of all probability assignments over latent variables.

3.4   Bayesian Inference and Beyond

With these last specifications, we are ready to apply the machinery of Bayesian statistical inference. We have a model, namely the space of multinomial distributions over observable and latent variables, suitably restricted by Equations (2.1) to (2.6). And we have a prior probability over this model. So from a sample of subjects with their scores on the observable variables, we can derive a posterior probability distribution over the possible multinomial distributions, which entails expectations for the latent variables and test scores of as yet unobserved subjects. This completes the exposition of a Bayesian statistical inference for the psychometric example.

But can we accommodate this full Bayesian inference in the progicnet framework? Recall that this framework only takes finite numbers of probability assignments as input. However, the space of multinomial distributions used in the foregoing comprises of a continuum of statistical hypotheses. Fortunately, this can be solved by making the θ-parameters of the above vary discretely, exactly like we made the hypotheses H_j on processing speed vary discretely in order to fit it into the progicnet framework. With this discretisation of the probability space, we can indeed accommodate the advanced version of Bayesian statistical inference in the progicnet framework, and use the common calculus to the inference problems.

There are, however, shortcomings of the Bayesian approach that invite us to supplement it with other approaches. It depends on the details of the relations between latent and observable variables whether inferences such as the above can guide us to a unique probability assignment over latent variables. As repeatedly indicated in the foregoing, different points in the space of multinomial distributions may have the same marginal likelihoods for the observable variables, and in such cases the statistical model is simply not identified. For example, setting aside the extreme cases, there will always be several probability assignments over the latent variables h_j^n that have maximal likelihood for the observed relative frequency of c_j^1. Unfortunately, this paper is too short to include a discussion of the exact conditions under which this occurs. But we are sure that if it does occur, the results of the statistical analysis crucially depend on the prior probability assignment over the model, and in a way that cannot be resolved by collecting more data.

Shortcomings of this kind call for different approaches to the problem presented by the psychometric example. To improve on the estimations we might, for example, try and employ statistical knowledge on test and latent variables for slightly different classes of subjects. In the next section we will show how evidential probability enables us to employ such knowledge, and furthermore how this approach is covered by the progicnet framework. Alternatively, we might try and avoid the use of priors over the model altogether and simply work with the set of probability assignments determined by the input. This is the approach of probabilistic argumentation, which is dealt with in Section 5. Finally, we may also take the preferred element in the set of allowed distributions under some preference ordering of probability distributions. This objective Bayesian approach, finally, is dealt with in Section 6.

4   Evidential Probability

The first of the above suggestions is nicely accommodated by evidential probability (EP). We will first briefly review EP and then illustrate it in the context of the psychometric example.

4.1   Introduction into EP

The theory of evidential probability rests on two central ideas [10, 12, 7]: probability assessments should be based upon relative frequencies, to the extent that we know them, and the assignment of probability to specific events should be determined by everything that is known about that event.

The crux of the difference between evidential probability and Bayesian statistical inference is how approximate joint statistical distributions are handled. Bayesian statistical methods assume that there are always joint distributions available for use, whereas evidential probability does not. Instead, EP maintains that there must be empirical grounds for assigning a joint frequency probability and that we must accept the uncertainty that attends our incomplete knowledge of statistical regularities. There are of course many inference problems where the two approaches perfectly align: both theories agree that Bayes's theorem is a theorem. But the two accounts differ sharply in their assessment of the range of reasonable applications of Bayesian inference structures, and whether the alternative evidential probability methods are appropriate. See Seidenfeld [14] and Kyburg [11] for a succinct comparison.

Evidential probability is conditional in the sense that the probability of a sentence ψ is relative to a finite set of sentences Γ_δ, which represent background knowledge. The evidential probability of $\psi(j)$ given Γ_δ, is an interval, $[l, u]$, in view of our limited knowledge of relative frequencies: $\mathrm{Prob}(\psi(t), \Gamma_\delta) = [l, u]$ expresses that the evidential probability that individual j is a ψ given the relevant statistical information in Γ_δ is $[l, u]$, where relevant information in Γ_δ *includes*

- the relative frequency information that the proportion of a reference set R that is also $\psi(j)$ is between l and u percent, and

- the information that the individual j is a member of R,

but *excludes*

- the relative frequency information of rival reference sets R^* to which j belongs that are no stronger than R, and

- all other frequency information about ψ *except* those from sets R' that j belongs to that are larger than R, i.e., $R \subset R'$.

There may well be several classes that satisfy these conditions with respect to $\psi(j)$, each with conflicting statistics to associate to j, but there is nevertheless a unique evidential probability assigned to $\psi(j)$ given Γ_δ: it is the smallest cover of the intervals associated with the set of undominated reference formulas.

Evidential probability is conditional in the sense that the probability of a sentence ψ is relative to a finite set of sentences Γ_δ, which represent background knowledge. The evidential probability of $\psi(j)$ given Γ_δ, written as $\mathrm{Prob}(\psi(j), \Gamma_\delta)$, is an interval, $[l, u]$, in view of our limited knowledge of relative frequencies: $\mathrm{Prob}(\psi(j), \Gamma_\delta) = [l, u]$ expresses that the evidential probability that individual j is a ψ given the relevant statistical information in Γ_δ is $[l, u]$, where relevant information in Γ_δ *includes*

- the relative frequency information that the proportion of a reference set R that is also $\psi(j)$ is between l and u, and

- the information that the individual j is a member of R,

but *excludes*

- the relative frequency information of rival reference sets R^* to which j belongs that are no stronger than R, and

- all other frequency information about ψ *except* those from sets R that j belongs to that are more specific than R', i.e., $R \subset R'$.

There may well be several classes that satisfy these conditions with respect to $\psi(j)$, each with conflicting statistics to associate to j, but there is nevertheless a unique evidential probability assigned to $\psi(j)$ given Γ_δ: it is the smallest cover of the intervals associated with the set of undominated reference formulas.

EP is much less easily accommodated in the progicnet framework than other semantics we consider, because EP employs probability distributions that are defined over different populations and the semantics for the entailment relation are determined primarily by rules for resolving conflict among relevant reference statistics. However, as is further worked out in the progicnet programme, the error probabilities that are associated with this type of inference can still be treated within in the progicnet framework.

4.2   Illustration in the Psychometric Example

Since all probability assessments in EP are based upon observed relative frequencies, the probabilistic components of our psychological theory—relations (2.5) and (2.6)—do not have direct expression within EP: there is no place for a 'latent' random variable within the theory. Nevertheless, the sentences representing the psychological theory within EP may include the bi-conditionals

$$f_j \leftrightarrow \rho$$
$$g_j \leftrightarrow \rho'$$
$$h_j \leftrightarrow \rho''$$

for all j, where each ρ is an open reference formula occurring in some or another closed *direct inference statement* in Γ_δ that effect the constraints described by (2.5) and (2.6). Direct inference statements are statements that record the upper (u) and lower (l) bounds of the frequency of items ($\vec{x}$) satisfying a specified reference class (ρ) that also satisfy a particular target class (τ), which are written in the form $\%\vec{x}(\tau(\vec{x}), \rho(\vec{x})) = [l, u]$. There may be several direct inference statements in Γ_δ in which each open reference formula appears, of course. We are simply specifying the potential statistics for our inference problem, and pointing out that the list of potential statistics are determined by knowledge in Γ_δ.

Suppose that we have a particular subject, $j = 5$. We said at the outset that EP uses two sources of knowledge for assigning probabilities that concern subject 5: it draws upon knowledge of relevant statistical regularities known to affect subject 5, and it draws upon everything that is known about *that* individual, subject 5. We now demonstrate how each of these features is exercised in EP, and how this is represented in terms of the fundamental question of the progicnet framework.

Imagine that we have the medical files on our subjects and that what warrants accepting constraint (2.5) is that none of them have a record of adverse exposure to lead during childhood, which is taken to be a quantity greater than 10 micrograms of lead per deciliter of blood. However, news reaches us that any exposure to lead greater than 5 micrograms per deciliter is adverse, and a review of files reveals that there are subjects in the study who have had exposure above this threshold. Thus a new parameter is introduced, D, for exposure to lead.

Our theory says that adverse exposure to lead reduces the pass rates for task B of late development subjects. In other words, (2.5) is now available in leaded (d) or unleaded ($\neg d$) grades:

$$\%j(b_j, \rho'_j \wedge h^n_j \wedge d_j) = \frac{n - m}{N}, \text{ for some positive } m < n \qquad (4.1)$$

$$\%j(b_j, \rho'_j \wedge h^n_j \wedge \neg d_j) = \frac{n}{N} \qquad (4.2)$$

So if we know that subject 5 was a late development subject exposed to lead as a child, we would discount his expected performance category H by m in predicting his success at task B, and if we know all this about subject 5 but that he was not poisoned as a child then we would predict his success at B to be $\frac{n}{N}$.

And what if we had no pediatric records for subject 5? Here we would expect a prediction of success on B to be within the interval $[\frac{n-m}{N}, \frac{n}{N}]$, since leaded and unleaded are values of a binary variable and thus represent mutually exclusive categories. Still we do not know which state subject 5 is in, and it won't do to pick some point in between: subject 5 is either a leaded or unleaded subject. Thus, the evidential probability assigned to the direct inference b_5 given that (4.1) and (4.2) are in Γ_δ, and that *no other relevant statistics* are known, is the interval $[\frac{n-m}{N}, \frac{n}{N}]$.

Suppose now that we want to know the developmental category G that subject 5 belongs to, and that Γ_δ is fixed. We know that there are replacements for (2.7) and (2.8) in Γ_δ, of the form

$$\%j(\rho', b_j, [0.2, 0.4]), \qquad (4.3)$$

$$\%j(\rho', c_j, [0.3, 0.5]) \qquad (4.4)$$

respectively. Sentence (4.3) expresses that a proportion between 0.2 and 0.4 of the subjects who pass B belong to observable class ρ', which has the same truth value as category 1 of G. Sentence (4.4) expressed that between 0.3 and 0.5 of the subjects who pass C also belong to observable class ρ, which has the same truth value in our theory as category 1 of G. Suppose subject 5 has passed B and has also passed C. What is the probability that he is in category 1? Subject 5 belongs to two references sets, B and C, that yield conflicting probabilities regarding subject 5's membership to category 1 of G. There are no reference sets to which j belongs that offer stronger frequency information, nor are there larger sets to which either B or C belong. Thus, B and C represent undominated relevant reference statistics for ρ'. Therefore, EP assigns the shortest cover to ρ', $[0.2, 0.5]$. Thus $\text{Prob}(g(j), \Gamma_\delta) = [0.2, 0.5]$.

Each of these inferences may be represented as an instance of the basic question,

$$\varphi_1^{X_1}, \ldots, \varphi_n^{X_n} \approx \psi^Y,$$

by substituting $\varphi_1^{X_1}, \ldots, \varphi_n^{X_n}$ by Γ_δ on the left hand side and ψ by an ordered pair, $\langle \chi, [l, u] \rangle$, on the right hand side, which expresses that the evidential probability of formula χ is $[l, u]$. So, the inference towards

$$\text{Prob}(g(j), \Gamma_\delta) = [0.2, 0.5]$$

would be represented as

$$\bigwedge_i \ulcorner \%x(\tau(x), \rho(x), [l', u'])_i^1 \urcorner \bigwedge_j \varphi_j^1 \approx \langle g(j), [0.2, 0.5] \rangle^1,$$

where the left hand side consists of the conjunction of all direct inference statements ($\ulcorner \%x(\tau(x), \rho(x), [l, u])^1 \urcorner$) and all logical knowledge about relationships between classes (φ^1), the entailment relation $\approx$ is non-monotonic, the right hand side asserts that the target sentence $g(j)$ is assigned $[0.2, 0.5]$. That $g(j)$ is $[0.2, 0.5]$ just means that the proportion of EP models of

$$\bigwedge_i \ulcorner \%x(\tau(x), \rho(x), [l', u'])_i^1 \urcorner \bigwedge_j \varphi_j^1$$

that also satisfy $g(j)$ is between $[0.2, 0.5]$. Since the semantics for $\approx$ are given by the rules for resolving conflict rather than by probabilistic coherence, we assign 1 to all premises and also to $\psi = \langle g(j), [0.2, 0.5] \rangle$.

This shows that EP fits into the progicnet framework. For statistical information that is fully certain the application of the common calculus is uninteresting, since the semantics for $\approx$ is determined by the EP rules for resolving conflicts among reference statistics. Nevertheless, we can pose a question about the robustness of an EP inference, where error probabilities are assigned to the statistical premises. This 'second-order' EP inference does utilize the calculus, and we refer to the joint progicnet paper [6] for details.

5   Probabilistic Argumentation

In the above we have concentrated on statistical questions concerning the psychometric example. Probabilistic argumentation tackles a different set of questions that we might ask about subjects and psychological attributes, concerning the logical relations between the attributes. To some extend such logical relations can be accommodated by Bayesian statistical inference, as was illustrated in Section 3. But probabilistic argumentation provides tools for dealing with logical and probabilistic relations without taking recourse to prior probability assignments.

5.1 Introduction into Probabilistic Argumentation

In the theory of probabilistic argumentation [3, 4, 5, 9], the available knowledge is partly encoded as a set of logical premises Φ and partly as a fully specified probability space $(\Omega, 2^\Omega, P)$. Variables which constitute the multivariate state space Ω are called *probabilistic*. This setting gets particularly interesting when some of the logical premises include *non-probabilistic* variables, i.e., variables that are not contained in the probability space. The two classical questions of the probability and the logical deducibility of a hypothesis ψ can then be replaced by the more general question of the probability of a hypothesis being logically deducible from the premises. In other words, we use the given logical constraints to carry the probability measure P from Ω into the state space of all variables involved.

For this, the state space Ω is divided into an area $\text{Args}(\psi) = \{\omega \in \Omega : \Phi_\omega \models \psi\}$ of so-called *arguments*, whose elements are each sufficient to make the hypothesis ψ a logical consequence of the premises, and another area $\text{Args}(\neg\psi) = \{\omega \in \Omega : \Phi_\omega \models \neg\psi\}$ of so-called *counter-arguments*, whose elements are each sufficient to make the complementary hypothesis $\neg\psi$ a logical consequence of the premises (by Φ_ω we denote the set of premises obtained from instantiating the probabilistic variables in Φ according to ω). Note that the premises themselves may restrict the possible states in the probability space, and thus serves as evidence to turn the given prior probability measure P into a (conditional) posterior probability measure P'.

The so-called *degree of support* of ψ is then the posterior probability of the event $\text{Args}(\psi)$,

$$\text{dsp}(\psi) = P'(\text{Args}(\psi)) = \frac{P(\text{Args}(\psi)) - P(\text{Args}(\bot))}{1 - P(\text{Args}(\bot))}, \qquad (5.1)$$

and its dual counterpart, the so-called *degree of possibility* of ψ, is 1 minus the posterior probability of the event $\text{Args}(\neg\psi)$,

$$\text{dps}(\psi) = 1 - P'(\text{Args}(\neg\psi)) = 1 - \text{dsp}(\neg\psi). \qquad (5.2)$$

Intuitively, degrees of support measure the presence of evidence supporting the hypothesis, whereas degrees of possibility measure the absence of evidence refuting the hypothesis. Probabilistic argumentation is thus concerned with probabilities of a particular type of event of the form *"the hypothesis is deducible"* rather than *"the hypothesis is true"*. Apart from that, they are classical additive probabilities in the sense of Kolmogorov's axioms. In principle, degrees of support and possibility can therefore be accommodated in the progicnet framework.

When it comes to quantitatively evaluate the truth of a hypothesis ψ, it is possible to interpret degrees of support and degrees of possibility as respective lower and upper bounds of an interval. The fact that such bounds are obtained without effectively dealing with probability intervals or probability sets distinguishes the theory from most other approaches to probabilistic logic. Note that the use of probability intervals or sets of probabilities is by no means excluded in the context of probabilistic argumentation. This would simply lead to respective intervals or sets of degrees of support and degrees of possibility. Indeed, in order to solve the psychometrical example from Section 2.1, it turns out that we need to introduce such intervals of support and possibility.

5.2   Illustration in the Psychometric Example

Looking at the example from Section 2.1 from the probabilistic argumentation perspective, we first observe that the probabilistic constraints (2.5) to (2.8) affect the variables B, C, G, and H only, whereas variables A and F are tied to variable G by (2.3) and (2.4) on a purely logical basis. This allows us to consider a set of premises $\Phi = \{f \wedge g \rightarrow \neg a, \neg g \rightarrow a\}$ and a restricted state space Ω which includes the variables B, C, G, and H, but not A and F. If further logical constraints are observed, for example $(a \wedge g) \vee b$ from (2.9) or any other complex formula, they can be easily incorporated by extending Φ accordingly. The multi-faceted psychometric example is thus a nice illustration of the setting on which probabilistic argumentation operates. It also underlines the large variety of inferential problems the progicnet framework accommodates.

Since the probabilistic constraints in the example do not sufficiently restrict the possible probability measures relative to Ω to a single function P, we must cope with a whole set $\mathbb{P}$ of such probability measures. Recall that we specified this set in Section 3, where we identified the space of multinomial distributions that is consistent with the relations provided in the psychometric example, Equations (3.7) to (3.11). Recall further that for Bayesian inference, even when it came to inference about a single subject, we needed to define a prior probability over the model. But probabilistic argumentation does not need any such prior. Relative to what we have already learnt about a subject, for example that she passed test A, each $P \in \mathbb{P}$ in the remaining set of probability assignments leads to respective degrees of support and possibility for a given hypothesis, for example the hypothesis that the subject passes test C.

Moreover, from the fact that all given probabilistic constraints are either point-valued or intervals, we know that the resulting sets for degrees of support and possibility will also be point-valued or intervals. Note that hypotheses involving only probabilistic variables B, C, G, or H have equal degrees of support and possibility, i.e., the two intervals will coincide in those cases, but this does not hold for hypotheses involving A or F. In general, we may interpret the numerical difference between respective degrees of support and possibility as a quantification of the amount of available evidence that is relevant to the hypothesis in question. Besides the usual interpretation of probabilities as additive degrees of belief, which is central to the Bayesian account of rational decision making, classical Bayesian inference is not designated to provide such a separate notion of *evidential strength* relative to the resulting degrees of belief.

From a computational point of view, however, the step from a fixed probability measure to a set of probability measure, as required in our example, makes the inferential procedure of probabilistic argumentation much more challenging. As suggested in Subsection 1.2, one solution would be to incorporate the given constraints over the probabilistic variables into a credal network [2], and to use that network to compute lower and upper probabilities for the events $\text{Args}(\psi)$ and $\text{Args}(\neg\psi)$ to finally obtain respective bounds for degrees of support and possibility. Thus, the progicnet framework neatly accommodates inferences in probabilistic argumentation that employ interval-valued degrees of support and possibility (for corresponding algorithms and technical technical details we refer to [6]).

As inference in credal networks still gets extremely costly, even for small or mid-sized networks, the solution sketched above is not always a satisfactory way out. More promising is the idea of choosing (according to some principles) the "best" probability measure among the ones in $\mathbb{P}$, and then proceed as in the default case. The next section proposes a possible strategy for this.

6   Objective Bayesianism

To some extent the previous sections have had the idea of the progicnet framework as an epistemological scheme in the background: the inferences in the psychometric example tell us what to believe on the basis of the input provided. In objective Bayesianism, this perspective is brought to the fore. To answer the questions posed at the end of Section 2.1, they are recast explicitly in terms of the strengths of one's beliefs. For example, given background knowledge, assumptions and data—such as Equations (2.1) to (2.6)—and the observed performance of a subject on tests A and B, how strongly should one believe that the subject will pass test C? By reformulating the questions this way, one can invoke the machinery of Bayesian epistemology.

6.1   Bayesian Epistemology and Objective Bayesianism

According to the Bayesian view of epistemology, the strengths of our beliefs should be representable by real numbers in the unit interval, and these numbers should satisfy the axioms of probability: an agent should believe a tautology to degree 1 and her degree of belief in a disjunction of mutually exclusive propositions should equal the sum of her degrees of beliefs in those individual propositions. Thus the strengths of the agent's beliefs should be representable by a probability function P. Moreover, an agent's degrees of belief should be compatible with her background knowledge, assumptions, data and evidence (which we shall collectively call her *epistemic background* or simply *evidence* $\mathcal{E}$). The notion of compatibility can be explicated by principles of the following kind:

1. If a proposition is in her evidence, then the agent should fully believe it.

2. The agent's degrees of belief should match her best estimates of the physical probabilities: if the agent knows that 70% of subjects who pass A and B also pass C, and she knows that the subject in question has passed A and B, but no other relevant facts, then she should believe that the subject will pass C to degree 0.7.

3. If no probability function fits the evidence using the above principles—the evidence is inconsistent—then some consistency maintenance strategy should be invoked. E.g., deem a probability function to be compatible with the evidence if it is compatible with a maximal consistent subset of the evidence.

4. If two probability functions are compatible with the evidence then so is any function that lies between them; if a sequence of probability functions are compatible with the evidence then so is the limit of that sequence.

Via principles 1 and 2 the evidence $\mathcal{E}$ imposes constraints χ on the agent's degrees of belief. The set of probability functions that satisfy these constraints will be denoted by $\mathbb{P}_\chi$. If this set is empty we may need to consider a set $\mathbb{P}'_\chi$ that is obtained by a consistency maintenance procedure (principle 3). Invoking principle 4 we consider the convex closure $[\mathbb{P}'_\chi]$ of this set of probability functions. Then $\mathbb{E}$, the set of probability functions that are compatible with the evidence $\mathcal{E}$, is just $[\mathbb{P}'_\chi]$. See [15, §5.3] for a more detailed discussion of these principles and their motivation.

Subjective Bayesian epistemology holds that an agent should set her degrees of belief according to any probability function in $\mathbb{E}$—she can subjectively choose which function to follow. Objective Bayesian epistemology, on the other hand, holds that while an agent's degrees of belief should be compatible with her evidence, her degrees of belief should equivocate on issues that are not decided by this evidence. Thus the agent's degrees of belief should be set according to a function $\mathbb{P}_\mathcal{E}$ in $\mathbb{E}$ that is maximally equivocal. Where the domain is specified by a finite set Ω of elementary outcomes, the function in $\mathbb{E}$ that is maximally equivocal is the function in $\mathbb{E}$ that is closest to function $P_=$ which gives the same probability $1/|\Omega|$ to each elementary outcome. ($P_=$ is called the *equivocator.*) Distance from the equivocator is measured by cross entropy $d(P, P_=) = \sum_{\omega \in \Omega} P(\omega) \log P(\omega)/P_=(\omega) = \sum_{\omega \in \Omega} P(\omega) \log(|\Omega| P(\omega))$. Distance from the equivocator is minimised when entropy $- \sum_{\omega \in \Omega} P(\omega) \log P(\omega)$ is maximised, and so this procedure is often called the *Maximum Entropy Principle* or *maxent* for short. On a finite domain, there will be a unique function $P_{\mathcal{E}}$ that is closest to $P_=$ in $\mathbb{E}$, so the agent has no choice about what degrees of belief to adopt—they are objectively determined by her evidence. (On an infinite domain—such as that determined by an infinite predicate language—there are cases in which degrees of belief are not objectively determined; nevertheless, $P_{\mathcal{E}}$ tends to be very highly constrained, leaving little room for subjective choice.)

Note that this equivocation requirement yields a substantial difference between subjective and objective Bayesian epistemology. If a doctor knows nothing about a particular patient, she is perfectly entitled, on the subjective Bayesian account, to fully believe that the patient does not have particular ailment A. On the objective Bayesian account, however, the doctor should equivocate—i.e., she should believe that the patient has A to degree $\frac{1}{2}$. This equivocation constraint is motivated by considerations of risk. More extreme degrees of belief tend to be associated with riskier actions: with a full belief in $\neg A$ the doctor is likely to dismiss the patient, who may then deteriorate or perish, but with degree of belief $\frac{1}{2}$ the doctor is likely to seek further evidence. Now one should not take on more risk than the evidence demands: if the evidence forces a full belief then so be it; if not, it would be rash to adopt a full belief. Thus one should equivocate as far as evidence allows. This line of argument is developed in [17].

The objective Bayesian approach fits into the progicnet programme as follows. First, objective Bayesian epistemology provides a semantics for the probabilistic logic framework of Schema (1.1): $\varphi_1^{X_1}, \ldots, \varphi_n^{X_n} \approx \psi^Y$. According to this semantics, the premisses $\varphi_1^{X_1}, \ldots, \varphi_n^{X_n}$ are construed as characterising the agent's evidence $\mathcal{E}$. Here $\varphi_i^{X_i}$ is understood as saying that the physical probability of φ_i is in X_i (perhaps as determined by appropriate frequency information). This evidence imposes constraints χ on an agent's degrees of belief, where $\chi = \{P(\varphi_1) \in X_1, \ldots, P(\varphi_n) \in X_n\}$. The set of probability functions compatible with this evidence is $\mathbb{E} = [\mathbb{P}'_\chi]$. An agent with this evidence should adopt degrees of belief represented by a function $P_{\mathcal{E}}$ in $\mathbb{E}$ that is maximally equivocal. The question arises as to what value $P_{\mathcal{E}}$ gives to ψ, and one can take $Y = \{P_{\mathcal{E}}(\psi) : P_{\mathcal{E}} \in \mathbb{E} \text{ is maximally equivocal}\}$. On a finite domain Y will be a singleton. Thus objective Bayesianism provides a natural semantics for Schema (1.1). Now according to the progicnet programme, probabilistic networks might be used to calculate Y. Indeed, as we shall now see, *objective Bayesian nets* can be used to calculate Y.

6.2   Illustration in the Psychometric Example

Returning to the psychometric case study, the objective Bayesian approach
provides the following recipe. Equations (2.1) to (2.6) and the subject's
performance on tests A and B constitute the evidence $\mathcal{E}$. We should then
believe that the subject will pass C to degree $\mathrm{P}_{\mathcal{E}}(C)$, where $\mathrm{P}_{\mathcal{E}}$ is the maximally equivocal probability function out of all those that are compatible
with $\mathcal{E}$.

In general, *objective Bayesian nets* can be used to calculate objective
Bayesian probabilities [16] and [15, §§5.6–5.8]. The idea here is that the objective Bayesian probability function $\mathrm{P}_{\mathcal{E}}$ can be represented by a Bayesian
net, now called an objective Bayesian net, and standard Bayesian network
algorithms can be invoked to calculate the required probabilities, such as
$\mathrm{P}_{\mathcal{E}}(C)$. Because this probability function is a maximum entropy probability function it will automatically satisfy certain probabilistic independencies
and the graph in the Bayesian network that represents these independencies
is rather straightforward to construct. Join two variables by an undirected
edge if they occur in the same constraint of $\mathcal{E}$. Then separation in the
resulting undirected graph implies independence in $\mathrm{P}_{\mathcal{E}}$: if X separates Y
from Z in the graph then it is a fact that $\mathrm{P}_{\mathcal{E}}$ renders Y and Z probabilistically independent conditional on X. This undirected graph can easily be
transformed into a directed acyclic graph that is required in a Bayesian net.

The example of Subsection 2.1 is actually a very special case. Here
Equation (2.1) is a consequence of the objective Bayesian procedure: since
there are no known connections between different subjects in $\mathcal{E}$, $\mathrm{P}_{\mathcal{E}}$ will
render the features of different subjects probabilistically independent. In
this example we also have a causal picture in the evidence, namely that
depicted in Figure 2, where the latent variables F, G and H are causes of
the test results. When we have a causal graph, the graph in the objective
Bayesian network is just this graph [15, §5.8], and hence the factorisation
of Equation (2.2) is also a consequence of the objective Bayesian procedure.
The evidence can thus be viewed as the causal graph Figure 2 together with
the constraints Equations (2.3) to (2.9). Since we have the graph in the
objective Bayesian net, it remains to determine the conditional probabil-
ity distributions, i.e., the distributions $\mathrm{P}_{\mathcal{E}}(F)$, $\mathrm{P}_{\mathcal{E}}(G)$, $\mathrm{P}_{\mathcal{E}}(H)$, $\mathrm{P}_{\mathcal{E}}(A|F,G)$,
$\mathrm{P}_{\mathcal{E}}(B|G,H)$, $\mathrm{P}_{\mathcal{E}}(C|H)$. Since the causal structure is known, these dis-
tributions can be determined iteratively: first determine the distribution
$\mathrm{P}_{\mathcal{E}}(F)$ that is maximally equivocal, then $\mathrm{P}_{\mathcal{E}}(G)$, and so on up to $\mathrm{P}_{\mathcal{E}}(C|H)$
[15, §5.8]. By iteratively maximising entropy we obtain:

$$
\begin{aligned}
&\mathrm{P}_{\mathcal{E}}(f) = 1/2, && \mathrm{P}_{\mathcal{E}}(a|f,g) = 0, && \mathrm{P}_{\mathcal{E}}(b|g,h^n) = n/N, \\
&\mathrm{P}_{\mathcal{E}}(g) = 1/2, && \mathrm{P}_{\mathcal{E}}(a|f,\neg g) = 1, && \mathrm{P}_{\mathcal{E}}(b|\neg g,h^n) = 0.4, \\
&\mathrm{P}_{\mathcal{E}}(h^n) = 1/N, && \mathrm{P}_{\mathcal{E}}(a|\neg f,g) = 1/2, && \mathrm{P}_{\mathcal{E}}(c|h^n) = (N+n)/2N, \\
& && \mathrm{P}_{\mathcal{E}}(a|\neg f,\neg g) = 1.
\end{aligned}
$$

With these probability distributions and the directed acyclic graph we have a Bayesian network and can use standard Bayesian network methods to answer probabilistic questions. For example, how strongly should we believe that subject j will pass C given that she has passed tests A and B?

$$P_\varepsilon(c_j|a_j, b_j) =$$

$$= \frac{\sum_{f_j, g_j, h_j} P_\varepsilon(c|h_j)P_\varepsilon(b|g_j, h_j)P_\varepsilon(a|f_j, g_j)P_\varepsilon(f_j)P_\varepsilon(g_j)P_\varepsilon(h_j)}{\sum_{f_j, g_j, h_j} P_\varepsilon(b|g_j, h_j)P_\varepsilon(a|f_j, g_j)P_\varepsilon(f_j)P_\varepsilon(g_j)P_\varepsilon(h_j)}$$

$$= \frac{\sum_{f_j, g_j, h_j} P_\varepsilon(c|h_j)P_\varepsilon(b|g_j, h_j)P_\varepsilon(a|f_j, g_j)}{\sum_{f_j, g_j, h_j} P_\varepsilon(b|g_j, h_j)P_\varepsilon(a|f_j, g_j)}$$

$$= \frac{24N(3N + 1) + (N + 1)(5N + 1)}{6N(21N + 5)} = 0.61 \text{ as } N \longrightarrow \infty.$$

With the more extensive evidence of Equations (2.1) to (2.9), the procedure is just the same, though of course the conditional distributions and final answer differ from those calculated above.

From a computational point of view, the objective Bayesian approach is relatively straightforward for two reasons. First, there is only a single probability function P_ε under consideration. As we have seen, other approaches deal with sets of probability functions. Second, since this function is obtained by maximising entropy, we get lots of independencies for free; these independencies permit the construction of a relatively sparse Bayesian net, which in turn permits relatively quick inferences.

Computationally feasibility is one reason for preferring the objective Bayesian approach over the Bayesian statistical methods of Section 3, but there are others. A second reason is that the whole approach is simpler under the objective Bayesian account: instead of defining (higher-order) probabilities over statistical models one only needs to define probabilities over the variables of the domain. It may be argued that the move to higher-order probabilities is only warranted when the evidence includes specific information about these higher-order probabilities. Such information is generally not available.

A third argument for preferring the objective Bayesian approach appeals to epistemological considerations. Since Bayesian statistics defines probabilities over statistical hypotheses, these probabilities must be interpreted epistemologically, in terms of degrees of belief—it makes little sense to talk of the chance or frequency of a statistical model being true. Hence the Bayesian statistical approach naturally goes hand in hand with Bayesian epistemology. Typically, Bayesian statisticians advocate a subjective Bayesian approach to epistemology—probabilities should fit the evidence but are otherwise a matter of subjective choice. As we have seen, however, there are good reasons for thinking that this is too lax: such an approach condones degrees of belief that are more extreme than the evidence warrants, and degrees of belief that are too extreme subject the believer to unjustifiable risks and so are irrational.

Hence Bayesian statistics should minimally be accompanied by a principled way to determine reasonable priors, such as is provided by objective Bayesian epistemology. While there is a growing movement of statisticians who advocate such a move, it is well recognised that objective Bayesian epistemology is much harder to implement on the uncountable domains of Bayesian statistics than the finite domain considered here. This is because there may be no natural equivocator on an uncountable domain (cf. the discussion of the wine-water paradox in [8]), unless we can provide an argument to favour a particular parameterisation of the domain.

For lack of a preferred parameterisation, we have a dilemma: Bayesian statistics needs to be accompanied by a Bayesian epistemology; if a subjective Bayesian epistemology is chosen then Bayesian statistics is flawed for normative reasons; on the other hand if an objective Bayesian epistemology is chosen then there are implementational difficulties; moreover, the move to higher-order probabilities should only be made where absolutely necessary. Such a move is not absolutely necessary in the example of this paper. It may be argued, therefore, that in the context of the case study considered here, the objective Bayesian approach outlined in this section is more appropriate than the Bayesian statistical approach of Section 3. Minimally, it will provide a valuable addition to the statistical treatment considered there.

7 Conclusion

In this paper we have sketched a number of different approaches to combining logical and probabilistic inference. We showed how each of these approaches can be used to answer questions in the context of a toy example from psychometrics, how each approach can be subsumed under a unifying framework, thereby making them amenable to a common underlying calculus. But what exactly did we gain in doing so? We give a number of reasons for saying that the formulation of framework and calculus, as part of an overarching progicnet programme, amounts to progress.

First of all, we hope to have shown that the standard statistical treatment of the psychometric example, in this case using Bayesian statistics, can be supplemented in various ways by other approaches to logical and probabilistic inference. The progicnet programme provides a way to unify these approaches systematically. More specifically, and as illustrated in the psychometric example, the progicnet framework allows us to supplement the statistical inference that is standard in the psychometric context with some powerful inference tools from logic, all subject to the same calculus. We believe that there are many cases, in the sciences and in machine learning, in which the context provides a lot of logical background knowledge. The psychometric example is one of them, but many more such examples can be found in data mining, bioinformatics, computational linguistics, and sociological modelling. In all of these fields the existing statistical techniques cannot optimally employ the logical background knowledge. The progicnet framework may provide the means to use logical and statistical background knowledge simultaneously, and in a variety of problem domains.

More specifically, let us reiterate the conclusions on the use of the different approaches, that were reached in the preceding sections.

Bayesian statistical inference allows for dealing with the standard inferential problems of the psychometric example. In this paper it serves as a backdrop against which the merits of the other approaches covered by the progicnet framework can be made precise. Note that this is not to say that Bayesian statistical inference occupies a central place in the progicnet framework more generally.

Evidential probability is particularly suited if we learn further statistical information that conflicts with the given statistical model or introduces further constraints on it. It provides us with the tools to incorporate this new information and find trade-offs, where Bayesian inference must remain silent.

Probabilistic argumentation can be employed to derive upper and lower bounds on the probability assignments on the basis of the statistical model and the logical relations between the variables in the model only, without presupposing any prior probability assignments. This is very useful for investigating the properties of the model and the probabilistic implications of logical relations.

Objective Bayesianism offers a principled technique for reducing a set of probability assignments, such as the statistical model of the example, to a single probability assignment. For complicated models with many parameters, this provides a powerful simplification, and thus efficient inferential procedures.

Other reasons for using a common framework are more internal to the philosophical debate. The field of probabilistic inference is rather disparate, and discussions over interpretation and applications frequently interfere with discussions to do with formalisation and validity. Perfectly valid inferences in one approach may appear invalid in another approach, and even while all approaches somehow employ Kolmogorov's measure theoretic notion of probability, what is being measured by probability, and consequently the treatment of probability in the approaches, varies wildly. We hope that by providing a common framework for probabilistic logic, we help to structure the discussions, and determine more clearly which disagreements are meaningful and which are not.

Finally, the existence of a common framework also proves useful on a more practical level. Now that we have described a common framework, we can apply the common calculus of credal networks to it. As indicated in Section 1, and roughly illustrated in Section 3 and 6, credal networks can play an important part in keeping inferences manageable in probabilistic logic. More generally, the application of these networks will lead to more efficient inferences within each of the approaches involved. We must admit, however, that in the confines of the present paper, we have not explained the advantages of using networks in detail. For the exact use of credal networks in the progicnet programme, we again refer the reader to the central progicnet paper [6].

References

[1] Vic Barnett. *Comparative Statistical Inference*. John Wiley, 1999.

[2] Fabio G. Cozman. Credal Networks. *Artificial Intelligence*, 120(2):199–233, 2000.

[3] Rolf Haenni. Towards a Unifying Theory of Logical and Probabilistic Reasoning. In Fabio Cozman, Robert Nau, and Teddy Seidenfeld, editors, *ISIPTA'05, 4th International Symposium on Imprecise Probabilities and Their Applications*, pages 193–202, 2005.

[4] Rolf Haenni. Probabilistic Argumentation. *Journal of Applied Logic*, 7:155–176, 2009.

[5] Rolf Haenni, Jürg Kohlas, and Norbert Lehmann. Probabilistic Argumentation Systems. In Dov M. Gabbay and Philippe Smets, editors, *Handbook of Defeasible Reasoning and Uncertainty Management Systems*, volume 5: Algorithms for Uncertainty and Defeasible Reasoning, pages 221–288. Kluwer Academic Publishers, 2000.

[6] Rolf Haenni, Jan-Willem Romeijn, Gregory Wheeler, and Jon Williamson. *Probabilistic Logic and Probabilistic Networks*. Springer-Verlag, to appear.

[7] William Harper and Gregory Wheeler, editors. *Probability and Inference: Essays In Honor of Henry E. Kyburg, Jr.*, volume 2 of *Texts in Philosophy*. College Publications, 2007.

[8] John Maynard Keynes. *A Treatise on Probability*. Macmillan, London, 1921.

[9] Jürg Kohlas. Probabilistic Argumentation Systems: A New Way to Combine Logic With Probability. *Journal of Applied Logic*, 1(3–4):225–253, 2003.

[10] Henry E. Kyburg, Jr. *Probability and the Logic of Rational Belief*. Wesleyan University Press, 1961.

[11] Henry E. Kyburg, Jr. Bayesian Inference with Evidential Probability. In William Harper and Gregory Wheeler, editors, *Probability and Inference: Essays in Honor of Henry E. Kyburg, Jr.*, volume 2 of *Texts in Philosophy*, pages 281–296. College Publications, 2007.

[12] Henry E. Kyburg, Jr. and Choh Man Tong. *Uncertain Inference*. Cambridge University Press, 2001.

[13] James Press. *Subjective and Objective Bayesian Statistics: Principles, Models, and Applications*. John Wiley, 2003.

[14] Teddy Seidenfeld. Forbidden Fruit: When Epistemic Probability May Not Take a Bite of the Bayesian apple. In William Harper and Gregory Wheeler, editors, *Probability and Inference: Essays in Honor of Henry E. Kyburg, Jr.*, volume 2 of *Texts in Philosophy*, pages 267–279. College Publications, 2007.

[15] Jon Williamson. *Bayesian Nets and Causality: Philosophical and Computational Foundations*. Oxford University Press, 2005.

[16] Jon Williamson. Objective Bayesian Nets. In Sergei Artemov, Howard Barringer, Artur S. d'Avila Garcez, Luis C. Lamb, and John Woods, editors, *We Will Show Them! Essays in Honour of Dov Gabbay. Volume 2*, volume 2 of *Tributes*, pages 713–730. College Publications, 2005.

[17] Jon Williamson. Motivating Objective Bayesianism: from Empirical Constraints to Objective Probabilities. In William L. Harper and Gregory R. Wheeler, editors, *Probability and Inference: Essays in Honour of Henry E. Kyburg Jr.*, volume 2 of *Texts in Philosophy*, pages 151–179. College Publications, 2007.

Benedikt **Löwe**, Eric **Pacuit**, Jan-Willem **Romeijn** (*eds.*)
Foundations of the Formal Sciences VI
Reasoning about Probabilities and Probabilistic Reasoning

The Principle of Conformity and Spectrum Exchangeability

JÜRGEN LANDES, JEFF PARIS, ALENA VENCOVSKÁ*

School of Mathematics, The University of Manchester, Manchester M13 9PL, United Kingdom
E-mail: juergen_landes@yahoo.de, {jeff.paris,alena.vencovska}@manchester.ac.uk

1 Introduction

In the 1950's R. Carnap and others [1, 2, 3] (following unbeknownst in the earlier footsteps of W. E. Johnson [8]) developed a framework that became known as inductive logic.[1] They were interested in predicting, by purely logical or rational considerations, the outcome of a simple experiment after running the experiment a few times. For example, suppose that you are in a foreign city and cannot make sense of the street signs. You observe the first six cars going from right to left. What can you conclude about the probability of the, as yet not observed, seventh car also proceeding in a leftward direction? What probability should you assign to it in fact being a one way street?

This simple example is a special case of the general problem of characterizing induction:

> *Given that you know only φ what probability should you rationally assign to θ?*

where θ, φ are sentences of some suitable predicate language (to be specified shortly). What we are investigating in this paper is a *logical* approach based on assuming explicit principles of rationality in order to constrain, if not exactly specify, what probability assignments are permissible. Thus the

---

*The first author was supported by a MATHLOGAPS Research Studentship, MEST-CT-2004-504029, the third author by a UK Engineering and Physical Sciences Research Council (EPSRC) Research Assistantship. The authors would like to thank the referees for their useful comments.
[1]Not to be confused with inductive logic programming.

Received by the editors: 29 May 2007; 14 December 2007, 11 January 2008.
Accepted for publication: 1 February 2008.

burden of justifying, or not, induction is carried by the supposed 'rationality' of these assumed principles, a matter on which we will but briefly comment leaving it largely to the reader to judge, the main focus of this paper being on their consequences for the above problem. Indeed as explained in, for example, [13, pp. 737–738], this question can really be pared down to the special case:

$$\text{What probability function } w \text{ should you rationally} \qquad (1.1)$$
$$\text{adopt if you know nothing at all?}$$

After some initial success Carnap's programme of attacking this problem via considerations of pure logic, or rationality, was largely abandoned following the challenge to its practicality presented by N.Goodman's "grue" Paradox, [5, 6]. With some few exceptions, e.g., [7, 10], work along these lines has remained sporadic until relatively recently.

Up to Kemeny 's paper [9], written around 1954, Carnap's programme had very largely been carried out for purely unary languages and Kemeny himself in this paper (pp. 733–734) points to the next step being generalizing it to predicates of higher arity. However with the exception of [7, 10] and recent work on binary languages in [12, 13, 15, 16] this path has remained very largely untrodden.

These latter papers identified an apparently central principle, *Spectrum Exchangeability*, which was arguably justified on grounds of rationality and at the same time implied many of the other rationality principles which had previously been proposed, in particular a version of *Unary Conformity*, cf. [12, pp. 70–71] and [13, p. 739]. Whilst this work was carried out in the context of a purely binary language both these principles generalize to finite languages with predicates of arbitrary arity. It is the purpose of this paper to show that in that wider context Spectrum Exchangeability implies a particularly strong generalization of Unary Conformity, what we shall call the *Principle of Conformity*. Again then both these principles can be viewed as natural logical constraints to place on the assigning of probabilities in the central problem of induction stated earlier.

2   Notation and machinery

Let L be a language with finitely many predicate symbols, the constant symbols $a_1, a_2, a_3, \ldots$, no equality and no function or additional constant symbols. The intention here is that these individuals a_i are distinct and exhaust the universe. Let SL, FL and $QFSL$ denote the sets of sentences, formulae and quantifier free sentences of L respectively.

For simplicity we shall start by assuming that L contains just one 4-ary predicate, R, relaxing this assumption once the main ideas and theorems have been derived.

Definition 2.1. A function $w : SL \to [0, 1]$ is a *probability function* if the following three conditions hold for all $\theta, \varphi, \exists x \psi(x) \in SL$:

(P1)   If $\models \theta$ then $w(\theta) = 1$.

(P2)   If $\models \neg(\theta \wedge \varphi)$ then $w(\theta \vee \varphi) = w(\theta) + w(\varphi)$.

(P3) $w(\exists x \psi(x)) = \lim_{m \to \infty} w(\bigvee_{i=1}^{m} \psi(a_i))$.

From now on w etc. will always denote a probability function. Condition (P3) is referred to as *Gaifman's condition*. The following result from [4, Theorem 1] shows that it is essentially redundant and that as far as probability functions are concerned we only really need to consider their restriction to the QFSL:

Theorem 2.2. Any function $v : QFSL \to [0, 1]$ satisfying (P1) and (P2) extends uniquely to a probability function on SL.

The following facts about probability functions (cf., e.g., [14, p. 162]) will be assumed throughout:

Proposition 2.3. If w is a probability function on SL and $\varphi, \psi \in SL$, then:

1. $w(\neg \varphi) = 1 - w(\varphi)$.

2. If $\models \varphi$ then $w(\neg \varphi) = 0$.

3. If $\varphi \models \psi$ then $w(\varphi) \leq w(\psi)$.

4. $w(\varphi \vee \psi) = w(\varphi) + w(\psi) - w(\varphi \wedge \psi)$.

5. If $\models \varphi \leftrightarrow \psi$ then $w(\varphi) = w(\psi)$.

For distinct $b_1, b_2, \ldots, b_p$ coming from the a_i (a notation we shall adopt throughout) we define:

Definition 2.4. A *state description for* $b_1, b_2, \ldots, b_p$ is a formula in QFSL of the form:

$$\bigwedge_{\vec{b} \in \{b_1, \ldots, b_p\}^4} R^{\varepsilon_{\vec{b}}}(\vec{b}) \tag{2.1}$$

where $\varepsilon_{\vec{b}} \in \{0, 1\}$ and $R^{\varepsilon_{\vec{b}}}(\vec{b})$ is $R(\vec{b})$ if $\varepsilon_{\vec{b}} = 1$ and $\neg R(\vec{b})$ if $\varepsilon_{\vec{b}} = 0$.

Let $\mathrm{SD}(\vec{b})$ be the set of such state descriptions of $b_1, b_2, \ldots, b_p$. Notice that such a state description can be completely described by the $\varepsilon_{\vec{b}}$. So we get for this predicate R an array of dimension of the arity of R (i.e., 4) where all indices run between 1 and p with entries that are either zero or one. The next lemma follows directly from Proposition 2.3.

Lemma 2.5. For all $\varphi(b_1, \ldots, b_p) \in \mathsf{QFSL}$, the value of $w(\varphi)$ is completely determined by the values of w on $\mathrm{SD}(\vec{b})$.

We now introduce the central notion of the *spectrum* of a state description. Given $\alpha \in \mathrm{SD}(\vec{b})$, say

$$\alpha = \bigwedge_{\vec{b} \in \{b_1, \ldots, b_p\}^4} R^{\varepsilon_{\vec{b}}}(\vec{b}),$$

define an equivalence relation $\mathcal{S}(\alpha)$ on $\{1, \ldots, p\}$ by:

$$\mathcal{S}(\alpha)ij \iff \text{for all } m, k, r \in \{1, \ldots, p\}, \text{ we have } \varepsilon_{b_i b_k b_m b_r} = \varepsilon_{b_j b_k b_m b_r},$$
$$\varepsilon_{b_k b_i b_m b_r} = \varepsilon_{b_k b_j b_m b_r}, \; \varepsilon_{b_k b_m b_i b_r} = \varepsilon_{b_k b_m b_j b_r}, \text{ and}$$
$$\varepsilon_{b_k b_m b_r b_i} = \varepsilon_{b_k b_m b_r b_j}.$$

In other words i and j are equivalent if according to the information given by α, b_i and b_j are *indistinguishable*.

Let the equivalence classes of $\mathcal{S}(\alpha)$ be $S_1, \ldots, S_k$, where $|S_i| \geq |S_j|$ for $i < j$, a situation we denote by $\mathcal{S}(\alpha) = \langle S_1, S_2, \ldots, S_k \rangle$, and define the *spectrum* of α to be the vector $\vec{\mathcal{S}(\alpha)} = \langle |S_1|, \ldots, |S_k| \rangle$. In other words the spectrum of a state description is the tuple of sizes of equivalence classes of indistinguishable indices, such that the components of this vector are non-increasing. Let X denote the *set of all spectra* and set $X^p := \{\vec{x} \in X \mid \sum_i x_i = p\}$.

Notice that if α is the state description of $b_1, b_2, \ldots, b_p$ and β is the state description of $b_1, b_2, \ldots, b_q$ with $p < q$ and $\beta \models \alpha$ (in other words β extends the information given by α) then b_i, b_j (with $i, j \leq p$) may be indistinguishable according to α but not according to β, for example because β may include $R(a_i, a_i, a_i, a_{p+1})$ and $\neg R(a_j, a_i, a_i, a_{p+1})$.

3   Principles

In [12, p. 79] and [13, pp. 740-741], arguments are given for the rationality of the following key principle:

Principle of Spectrum Exchangeability (SX): Let $\vec{b}, \vec{c}$ have the same length and let $\alpha \in \mathrm{SD}(\vec{b})$ and $\beta \in \mathrm{SD}(\vec{c})$ be such that $\vec{\mathcal{S}(\alpha)} = \vec{\mathcal{S}(\beta)}$. Then $w(\alpha) = w(\beta)$.

In other words $w(\alpha)$ depends only on the spectrum of $\mathcal{S}(\alpha)$. Henceforth we shall assume that the probability functions on $\mathsf{S}L$ that we are considering satisfy (SX).[2] One consequence of this is that we can unequivocally write

---

[2]In the case of a purely unary language (SX) becomes *Atom Exchangeability*, see for example [14, Lemma 12.7], and as such, being a consequence of Johnson's Sufficientness Postulate, was acceptable to Johnson, Carnap, and others.

$w(\vec{x})$ for $\vec{x} \in X$ meaning $w(\alpha)$ for one/any state description α with spectrum $\vec{x}$.

In [12, 13] and the forthcoming [11] it is shown that the principle (SX) implies a number of other, well established, exchangeability principles, in particular *Constant Exchangeability* asserting that w is invariant under permutation of the constants a_i.

In the case of a binary language one somewhat more novel principle which was shown in [12, Theorem 5.19] and [13, Theorem 1] to follow from (SX) was *Unary Conformity*. It is the main purpose of this paper to show that a strong generalization of that principle, which we shall call the *Principle of Conformity*, also follows from (SX) for general finite predicate languages.

We are presently considering the simple case of L having just a single 4-ary predicate symbol. To motivate this principle in our simple case, let L_P be the corresponding language with a single 3-ary predicate symbol P and for $\theta \in SL_P$, let θ_1 be the result of replacing each occurrence of $P(t_1, t_2, t_3)$ (where the t_i are terms of L) in θ by $R(t_1, t_2, t_3, a_1)$. Similarly, let θ_2 be the result of replacing each occurrence of $P(t_1, t_2, t_3)$ in θ by $R(t_1, t_2, t_3, t_3)$. So now θ_1, θ_2 are sentences of L and have probabilities $w(\theta_1), w(\theta_2)$.

Now suppose that w is to be a putative solution to the aforementioned problem (1.1). Can we justify saying that $w(\theta_1)$ should be greater than $w(\theta_2)$, or vice-versa? We would claim that we cannot, that there is no rational reason for supposing that $w(\theta_1)$ should exceed $w(\theta_2)$ nor any reason for supposing that $w(\theta_2)$ should exceed $w(\theta_1)$. In turn then, any putative solution w to (1.1) which did give them different values would seem to be making an unjustified, even unjustifiable, assumption.[3] The choice w would however be immune to this criticism if *it gave them both the same value*. We now show in this special case that w satisfying (SX) is enough to ensure such equality. We shall then go on to explain the general *Principle of Conformity*, which again follows from (SX)[4], and of which the above is but a particular example. [The reader whose main interest is in the general principle of which it is an exemplar may wish to skip the proof at this point.]

Proposition 3.1. For θ_1, θ_2 and w satisfying (SX) as above,

$$w(\theta_1) = w(\theta_2).$$

---

[3]One might initially feel that the repeated substitution of the same term, as with t_3 in $R(t_1, t_2, t_3, t_3)$, should be somehow distinguishable from $R(t_1, t_2, t_3, a_1)$ where there is no such repeat. However that amounts to maintaining that for a predicate R about which one knows nothing, its restriction, say, to the 'diagonal' $R(a_2, a_3, x_1, x_1)$ should be of a fundamentally different nature to its restriction to the 'column' $R(a_2, a_3, x_1, a_1)$. We can see no reason why this should be. Certainly within this context a convincing counter argument to this view would seem to reveal an intriguing new facet to the general Inductive Logic programme in which we are engaged.

[4]Consequently this principle inherits any justification we may make for (SX).

Proof. First notice that it is enough to take θ to be a state description,

$$\bigwedge_{\vec{c} \in \{b_1,\ldots,b_p\}^3} P^{\varepsilon \vec{c}}(c_1, c_2, c_3)$$

where $\vec{c} = \langle c_1, c_2, c_3 \rangle$ and (in this example) $a_1 \in \{b_1, \ldots, b_p\}$. In this case then

$$\theta_1 = \bigwedge_{\vec{c} \in \{b_1,\ldots,b_p\}^3} R^{\varepsilon \vec{c}}(c_1, c_2, c_3, a_1),$$

$$\theta_2 = \bigwedge_{\vec{c} \in \{b_1,\ldots,b_p\}^3} R^{\varepsilon \vec{c}}(c_1, c_2, c_3, c_3).$$

The formulas θ_1, θ_2 are not themselves state descriptions but since w satisfies (SX) it is enough to show that for any spectrum $\vec{x} \in X^p$ the number of state descriptions for $b_1, b_2, \ldots, b_p$ extending θ_1 with spectrum $\vec{x}$ is the same as the number extending θ_2 with spectrum $\vec{x}$.

In fact we shall show something slightly stronger than this, namely that for $\sigma := \langle S_1, S_2, \ldots, S_k \rangle$ a partition of $\{1, 2, \ldots, p\}$ the sets

$$A_p(\sigma) = \{\alpha \in \mathrm{SD}(\vec{b}) \mid \alpha \text{ extends } \theta_1 \text{ and } \mathcal{S}(\alpha) = \sigma\}$$
$$B_p(\sigma) = \{\alpha \in \mathrm{SD}(\vec{b}) \mid \alpha \text{ extends } \theta_2 \text{ and } \mathcal{S}(\alpha) = \sigma\}$$

have the same size. Notice that for all $\alpha \in A_p(\sigma)$, $\mathcal{S}(\alpha)$ is a refinement of $\mathcal{S}(\theta)$, and similarly for $\alpha \in B_p(\sigma)$.

The proof proceeds by induction on k, the number of classes of σ, where we can assume that σ is a refinement of $\mathcal{S}(\theta)$. If k equals the length of $\mathcal{S}(\theta)$ then A_p and B_p have size $2^{k^3(k-1)}$ if $\mathcal{S}(\theta) = \langle S_1, \ldots, S_k \rangle$ and size 0 otherwise. To see this notice that for $1 \leq i, j, r, l \leq k$ any $\alpha \in A_p(\sigma)$ must force $R(c_1, c_2, c_3, c_4)$ to be simultaneously true/false for all $c_1 \in S_i, c_2 \in S_j, c_3 \in S_r, c_4 \in S_l$. For this there is a free choice unless $a_1 \in S_l$, when there is no choice. Hence the number of possible combinations of choices is $2^{k^3(k-1)}$. Similarly for $\alpha \in B_p(\sigma)$.

For the induction step, let $\sigma = \langle S_1, \ldots, S_k \rangle$ be a non-trivial refinement of $\mathcal{S}(\theta)$, denoted $\mathcal{S}(\theta) < \sigma$. We show that

$$\sum_{\mathcal{S}(\theta) \leq \tau \leq \sigma} |A_p(\tau)| = \sum_{\mathcal{S}(\theta) \leq \tau \leq \sigma} |B_p(\tau)| \ . \tag{3.1}$$

For $i \leq k$, let s_i be the smallest element in S_i. Consider $\alpha \in \mathrm{SD}(\vec{b})$ such that α extends θ_1 and $\mathcal{S}(\alpha) \leq \sigma$. Then, since the $s_1, \ldots, s_k$ contain representatives (not necessarily unique) for each of the equivalence classes in $\mathcal{S}(\alpha)$, α is determined by the $\gamma \in \mathrm{SD}(\{b_{s_1}, \ldots, b_{s_k}\})$ which it extends. Clearly also $\gamma \wedge \theta_1$ is consistent.

Conversely, given $\gamma \in \mathrm{SD}(\{b_{s_1}, \ldots, b_{s_k}\})$ such that $\gamma \wedge \theta_1$ is consistent we can define a unique $\alpha \in \mathrm{SD}(\bar{b})$ determined by γ as above such that $\mathcal{S}(\alpha) \leq \sigma$ (since then the $s_1, \ldots, s_k$ must contain representatives, not necessarily unique, for each of the equivalence classes in $\mathcal{S}(\alpha)$). Furthermore α will have to extend θ_1. To see this notice that if $R^{\varepsilon\bar{\varepsilon}}(c_1, c_2, c_3, a_1)$ is a conjunct in α then $R^{\varepsilon\bar{\varepsilon}}(b_{s_i}, b_{s_j}, b_{s_r}, b_{s_l})$ also appears in α, where $a_1 \in S_l$, $c_1 \in S_i$ etc., so this also appears in γ and hence in θ_1. But then $P^{\varepsilon\bar{\varepsilon}}(b_{s_i}, b_{s_j}, b_{s_r})$ appears in θ so since $\mathcal{S}(\theta) \leq \sigma$, $P^{\varepsilon\bar{\varepsilon}}(c_1, c_2, c_3)$ appears in θ and $R^{\varepsilon\bar{\varepsilon}}(c_1, c_2, c_3, a_1)$ appears in θ_1.

As above, it follows that

$$\sum_{\mathcal{S}(\theta) \leq \tau \leq \sigma} |A_p(\tau)| = |\{\gamma \in \mathrm{SD}(\{b_{s_1}, \ldots, b_{s_k}\}) \mid \gamma \wedge \theta_1 \text{ is consistent}\}|$$

$$- 2^{k^3(k-1)} \tag{3.2}$$

by observing that $|\mathrm{SD}(\{b_{s_1}, \ldots, b_{s_k}\})| = 2^{(k^4)}$ and we need to fix k^3 entries in the array for these γ to accommodate $\theta_1 \upharpoonright (b_{s_1}, \ldots, b_{s_k})$. *Mutatis mutandis,*

$$\sum_{\mathcal{S}(\theta) \leq \tau \leq \sigma} |B_p(\tau)| = |\{\gamma \in \mathrm{SD}(\{b_{s_1}, \ldots, b_{s_k}\}) \mid \gamma \wedge \theta_2 \text{ is consistent}\}|$$

$$= 2^{k^3(k-1)}. \tag{3.3}$$

So (3.1) is proved, and the induction hypothesis and (3.1) show that $|A_p(\sigma)| = |B_p(\sigma)|$ as required.   Q.E.D.

Notice that having obtained this equality in the case of $R(x_1, x_2, x_3, a_1)$ and $R(x_1, x_2, x_3, x_3)$ we can by analogous transpositions derive equality for the cases of

$$R(x_1, x_2, a_3, x_3), R(x_1, x_2, x_1, x_3), R(a_2, x_2, x_1, x_3), \ldots, R(t_1, t_2, t_3, t_4),$$

provided the variables amongst the terms t_1, t_2, t_3, t_4 are exactly x_1, x_2, x_3. So for example the Principle of Conformity will imply for all fixed $k, l \in \mathbb{N}$

$$w\left(\bigwedge_{i,j,n=1}^{p} R^{\varepsilon_{ijn}}(a_i, a_n, a_j, a_k)\right) = w\left(\bigwedge_{i,j,n=1}^{p} R^{\varepsilon_{ijn}}(a_l, a_i, a_j, a_n)\right) =$$

$$= w\left(\bigwedge_{i,j,n=1}^{p} R^{\varepsilon_{ijn}}(a_j, a_n, a_{p+1}, a_i)\right)$$

$$= w\left(\bigwedge_{i,j,n=1}^{p} R^{\varepsilon_{ijn}}(a_n, a_j, a_i, a_j)\right). \tag{3.4}$$

4   The Principle of Conformity

The generalization of the proof of Proposition 3.1 to larger languages involves no new ideas, the main problem is keeping the notation within acceptable limits. To this end suppose our language L has predicate symbols
$R_1, R_2, \ldots, R_h$ of arities $d_1, d_2, \ldots, d_h$ respectively. Let L^- be a language
with predicate symbols $P_1, P_2, \ldots, P_h$ with arities $e_1, e_2, \ldots, e_h$ such that
$e_j \leq d_j$ for $j = 1, 2, \ldots, h$.

For each $i = 1, 2, \ldots h$, let $u_1^i, u_2^i, \ldots, u_{d_i}^i$ and $s_1^i, s_2^i, \ldots, s_{d_i}^i$ be terms
of L such that the variables appearing in each of these lists are exactly
$x_1, x_2, \ldots, x_{e_i}$ and let

$$R_i^{\vec{u}}(x_1, x_2, \ldots, x_{e_i}) = R_i(u_1^i, u_2^i, \ldots, u_{d_i}^i), \text{ and}$$
$$R_i^{\vec{s}}(x_1, x_2, \ldots, x_{e_i}) = R_i(s_1^i, s_2^i, \ldots, s_{d_i}^i).$$

Finally, for $\theta \in SL^-$, let $\theta_1 \in SL$ be the result of replacing each
$P_i(t_1, t_2, \ldots, t_{e_i})$ in θ by $R_i^{\vec{u}}(t_1, t_2, \ldots, t_{e_i})$ and let $\theta_2 \in SL$ be the result
of replacing each $P_i(t_1, t_2, \ldots, t_{e_i})$ in θ by $R_i^{\vec{s}}(t_1, t_2, \ldots, t_{e_i})$.

Principle of Conformity (PC): Whenever the $R_i, P_i, u_j^i, s_j^i, \theta, etc.$ are as
above, $w(\theta_1) = w(\theta_2)$.

The intuition here is that in this context of assigning probabilities there
is no reason for w to treat the $R_i^{\vec{u}}(t_1, t_2, \ldots, t_{e_i})$ and $R_i^{\vec{s}}(t_1, t_2, \ldots, t_{e_i})$ in any
way differently, and so turn no reason for w to treat θ_1 and θ_2 differently.

Theorem 4.1. (SX) $\Rightarrow$ (PC).

All the key ideas for the proof of this theorem appear already in the
proof of Proposition 3.1. For the full details, cf. [11]. In particular (3.4) is
a special case of this theorem when $h = 1, d_1 = 4, e_1 = 3$, the variables are,
for instance, x_1, x_2, x_3, the terms for the first and last expressions are, for
example, $u_1^1 = x_1, u_2^1 = x_2, u_3^1 = x_3, u_4^1 = a_k$ and $s_1^1 = x_2, s_2^1 = x_3, s_3^1 = x_1, s_4^1 = x_3$.

5   Conclusion

The naturalness of the principle (SX) for binary languages has been argued
in [12, pp. 70–71] and [13, p. 739] and similar remarks apply here where
we are considering arbitrary arities. As was shown in those texts, and
continued here, this assumption (SX) has a number of attractive properties.
We would anticipate that further research will add still more and that (SX)
will become established as a central channel through which to investigate
inductive logic.

References

[1] Rudolf Carnap. *Logical Foundations of Probability*. University of Chicago Press, 1950. 2nd edition 1962.

[2] Rudolf Carnap. *The Continuum of Inductive Methods*. University of Chicago Press, 1952.

[3] Rudolf Carnap. A Basic System of Inductive Logic, Part II. In Richard C. Jeffrey, editor, *Studies in Inductive Logic and Probability*, volume 2, pages 1–155. University of California Press, 1980.

[4] H. Gaifman. Concerning Measures on First Order Calculi. *Israel Journal of Mathematics*, 2:1–18, 1964.

[5] Nelson Goodman. A Query on Confirmation. *Journal of Philosophy*, 43:383–385, 1946.

[6] Nelson Goodman. On Infirmities in Confirmation-Theory. *Philosophy and Phenomenology Research*, 8:149–151, 1947.

[7] Douglas N. Hoover. Relations on Probability Spaces and Arrays of Random Variables. Technical report, Institute for Advanced Study, Princeton, 1979.

[8] William Ernest Johnson. Probability: The Deductive and Inductive Problems. *Mind*, 41:409–423, 1932.

[9] John. G. Kemeny. Carnap's Theory of Probability and Induction. In Paul Arthur Schilpp, editor, *The Philosophy of Rudolf Carnap*, pages 711–738. Open Court, 1963.

[10] Peter Krauss. Representation of Symmetric Probability Models. *Journal of Symbolic Logic*, 34:183–193, 1969.

[11] Jürgen Landes. *The Principle of Spectrum Exchangeability within Inductive Logic*. PhD thesis, Manchester University, 2009.

[12] Chris Nix. *Probabilistic Induction in the Predicate Calculus*. PhD thesis, Manchester University, 2005.

[13] Chris Nix and Jeff B. Paris. A Note on Binary Inductive Logic. *Journal of Philosophical Logic*, 36:735–771, 2007.

[14] Jeff B. Paris. *The Uncertain Reasoner's Companion*. Cambridge University Press, 1994.

[15] Jeff B. Paris and Alena Vencovská. From Unary to Binary Inductive Logic, 2007. Presented at the *Second Indian Conference on Logic and Its Relationship with Other Disciplines*, IIT Bombay, January 2007.

[16] Alena Vencovská. Binary Induction and Carnap's Continuum. In Didier Dubois, Petr Hajek, Radim Jirousek, Gernot D. Kleiter, Peter Naeve, and Romano Scozzafava, editors, *Proceedings of the 7th Workshop on Uncertainty Processing (WUPES)*, pages 173–182, 2006.

Benedikt **Löwe**, Eric **Pacuit**, Jan-Willem **Romeijn** (*eds.*)
Foundations of the Formal Sciences VI
Reasoning about Probabilities and Probabilistic Reasoning

A Dutch Book for Group Decision-Making?

LUC BOVENS, WLODEK RABINOWICZ

[1] Department of Philosophy, Logic and Scientific Method, London School of Economics, Houghton Street, London, WC2A 2AE, United Kingdom

[2] Department of Philosophy, Lund Universitet, Kungshuset, Lundagård, 222 22 Lund, Sweden

E-mail: l.bovens@lse.ac.uk; Wlodek.Rabinowicz@fil.lu.se

1 The Problem of the Hats and a Dutch Book for a Group of Rational Players

The problem of the hats is a mathematical puzzle introduced in 1998 by Todd Ebert [4]. We randomly distribute white and black hats to a group of n rational players. We do this in the dark, with each player having an independent fifty-fifty chance of receiving a hat of one colour or the other. When the lights are turned on, each player can see the colour of the hats of the other players but not of his own hat. They are asked to simultaneously name the colour of their own hats or to pass. If at least one person is correct and nobody is in error, a prize will be awarded to the group. They can engage in pre-play communication (before the hats are distributed) to design a strategy. What is the optimal strategy? For $n = 3$, the solution is simple. If all and only the players who see two hats of the same colour call as the colour of their own hat the colour that is different from the colour of the two hats that they see, whereas the others pass, then the group wins in all cases in which the hats are not of the same colour — that is, in $\frac{3}{4}$ of the cases. It is an open problem, however, whether there exists a general solution, for an arbitrary n, to this optimisation problem.

We do not intend to tackle the problem of the hats, but take it as a starting point for setting up a Dutch Book, or what seems like a Dutch book, against a group of rational players with common priors and full trust in each other's rationality.

Received by the editors: 3 September 2007.
Accepted for publication: 15 November 2007.

Again distribute the hats amongst a group of three players. This time we disallow pre-play communication. Clearly, the chance that

$$\text{not all hats are of the same colour} \tag{A}$$

is $\frac{3}{4}$. The light is switched on and all players can see the colour of the hats of the other persons, but not the colour of their own hats. Then no matter what combination of hats was assigned, at least one player will see two hats of the same colour. For her, the chance that not all hats are the same colour strictly depends on the colour of her own hat and so equals $\frac{1}{2}$.

On Lewis's principal principle, a rational player will let her degrees of belief be determined by these chances. So before the light is switched on, all players will assign degree of belief of $\frac{3}{4}$ to (A) and after the light is turned on, at least one player will assign degree of belief of $\frac{1}{2}$ to (A). Suppose that before the light is turned on a bookie offers to sell a single bet on (A), with stakes \$4 at a price of \$3, and subsequently offers to buy a single bet on (A), with stakes \$4 at a price of \$2, after the light is turned on. Suppose, finally, that all of the above is common knowledge among the players.

If, following Ramsey, the degree of belief equals the betting rate (i.e., the price-stake ratio) at which the player is both willing to buy and willing to sell a bet on a given proposition, then any of the players would be willing to buy the first bet and at least one player would be willing to sell the second bet. Whether all hats are of the same colour or not, the bookie can make a Dutch book — she has a guaranteed profit of \$1:

	(A) is true: Not all hats are of same colour	(A) is false: All hats are of same colour
Bet 1	Player buys bet for \$3 Bookie pays out \$4	Player buys bet for \$3 Bet is lost
Bet 2	Bookie buys bet for \$2 Player pays out \$4	Bookie buys bet for \$2 Bet is lost
Payoffs	Bookie gains \$1	Bookie gains \$1

So, seemingly, the bookie has succeeded in making a Dutch book against a group of rational players. But the fact that a Dutch book can be made is a mark of some form of irrationality. There are two possibilities. Either each player is trying to increase her own payoff. Then the Dutch Book would not be too worrisome. After all, prisoner's dilemmas have a similar structure — when each player acts to her own advantage, the group payoff is suboptimal. In this case individual rationality would just not be in line with group rationality. Or, alternatively, the players are supposed to act in the interest of the group as a whole, i.e., to maximize the group's total payoff, rather than their own winnings. In this case the Dutch book would be worrisome — it would be indicative of an internal breakdown in group rationality.

In order not to keep the reader in suspense, let us say straightaway that the rational course of action is not to sell the bookie the second bet and hence that no Dutch book can be made — neither when the players are trying to increase their own payoffs nor when they are trying to increase the group payoff. But nonetheless, our solution to the paradox will prove rewarding. It forces us to think carefully about how one should state Ramsey's claim that degrees of belief are the betting rates at which a player is willing to buy or sell bets on a given proposition. We will conclude with suggestions how the reasoning in this hats puzzle is relevant to the tragedy of the commons and to strategic voting.

2 The Dutch Book Disarmed for Self-Interested Decision-Makers

Let us focus on the second bet — the one that the bookie offers to buy. Suppose the players are trying to maximise their own expected payoffs. Then we need to determine the probability p_i with which a player i should step forward and offer to sell the bet if she sees two hats of the same colour. (Obviously, she shouldn't step forward to sell the bet if she sees two hats of different colours. For in that case she knows that the bet would be won by the bookie.) Since there is no pre-play communication and the players are symmetrically placed, we are looking for a symmetrical Nash equilibrium $\langle p_a, p_b, p_c \rangle = \langle p, p, p \rangle$ for players Alice, Bob and Carol.

Suppose Alice sees two hats of the same colour. We calculate the expected utility for Alice of stepping forward given that she sees two hats of the same colour, $EU_a[\langle 1, p, p \rangle]$, by first conditioning on the random variable S with values $S = 0$ when the hats are of different colours and $S = 1$ when all the hats are of the same colour. Obviously, $P(S = 1) = 1 - P(S = 0)$.

$$\begin{aligned} EU_a[\langle 1, p, p \rangle] &= EU_a[\langle 1, p, p \rangle \mid S = 0]P(S = 0) \\ &+ EU_a[\langle 1, p, p \rangle \mid S = 1](1 - P(S = 0)). \end{aligned} \tag{1}$$

Since Alice has seen two hats of the same colour, $P(S = 0) = \frac{1}{2}$. If the hats are of different colours, Alice is the only one who will step forward and so her payoff $EU_a[\langle 1, p, p \rangle \mid S = 0] = -\$2$. (Remember that the bookie will win the second bet if $S = 0$.) To calculate $EU_a[\langle 1, p, p \rangle \mid S = 1]$, we condition on the random variable N with values $N = i$ when i other persons beside Alice decide to step forward to sell the bet, for i ranging over 0, 1 and 2. Hence,

$$\mathrm{EU}_a[\langle 1, p, p \rangle \mid S = 1] =$$
$$= \sum_{i=0,1,2} \mathrm{EU}_a[\langle 1, p, p \rangle \mid S = 1, N = i] \mathrm{P}(N = i \mid S = 1). \qquad (2)$$

The assumption is that the bookie chooses randomly among the players who offer to sell the bet he wants to buy. Thus, if i other players beside Alice step forward to sell the bet, each of them, Alice included, has the chance of $1/(i+1)$ of being the seller. The values of the components of the sum in the right-hand side of Equation (2) can therefore be read off from the following matrix (remember that the player who gets to sell the bet will win \$2 if $S = 1$):

i	$\mathrm{EU}_a[\langle 1, p, p \rangle \mid S = 1, N = i]$	$\mathrm{P}(N = i \mid S = 1)$
0	$2 \ (= 2 \times 1/1)$	$(1-p)^2$
1	$1 \ (= 2 \times \frac{1}{2})$	$2p\,(1-p)$
2	$2/3 \ (= 2 \times 1/3)$	p^2

Hence,

$$\mathrm{EU}_a[\langle 1, p, p \rangle] = \frac{(-2) + (2(1-p)^2 + 2p(1-p) + (2/3)p^2)}{2}$$
$$= 1/3(p-3)p. \qquad (3)$$

We now examine symmetric Nash equilibria in pure strategies. In view of Equation (3), unilateral deviation from $\langle 1,1,1 \rangle$ to $\langle 0,1,1 \rangle$ increases Alice's payoff from $\mathrm{EU}_a[\langle 1,1,1 \rangle] = 1/3(1-3)1 = -2/3$ to $\mathrm{EU}_a[\langle 0,1,1 \rangle] = 0$, but unilateral deviation from $\langle 0,0,0 \rangle$ to $\langle 1,0,0 \rangle$ leaves Alice's payoff constant at $\mathrm{EU}_a[\langle 0,0,0 \rangle] = \mathrm{EU}_a[\langle 1,0,0 \rangle] = 1/3(0-3)0 = 0$. Hence $\langle 0,0,0 \rangle$ is the only symmetric equilibrium in pure strategies. To examine whether there is an equilibrium in mixed strategy, we note that, if $0 < p < 1$, then $\langle p, p, p \rangle$ can be a Nash equilibrium only if the pure strategies that p is the mixture of have the same expected utility:

$$\mathrm{EU}_a[\langle 1, p, p \rangle] = \mathrm{EU}_a[\langle 0, p, p \rangle]$$
$$1/3(p-3)p = 0. \qquad (4)$$

However, Equation (4) has no solution under the constraint $0 < p < 1$. Hence, there exists only one symmetric equilibrium, viz. $\langle 0,0,0 \rangle$.

3   The Dutch Book Disarmed for Group-Interested Decision-Makers

Now suppose that each group member instead is out to maximize the payoff of the group. So what should a player do who sees two hats of the same colour?

Consider

$$
\begin{aligned}
\mathrm{EU}_g[\langle 0, p, p\rangle] \;=\;& \mathrm{EU}_g[\langle 0, p, p\rangle \mid S = 0]\mathrm{P}(S = 0) \\
+\;& \mathrm{EU}_g[\langle 0, p, p\rangle \mid S = 1](1 - \mathrm{P}(S = 0)). \quad (5)
\end{aligned}
$$

Once again, $\mathrm{P}(S = 0) = \frac{1}{2}$. (Note that it is Alice's probability that is in question, since it is her expectation of the group utility that we are after. Thus, it would be more appropriate to write $E_a U_g[\langle 0, p, p\rangle]$ instead of $\mathrm{EU}_g[\langle 0, p, p\rangle]$, but we omit the extra index to keep the notation simpler.) Furthermore, $\mathrm{EU}_g[\langle 0, p, p\rangle \mid S = 0] = 0$, since, if $S = 0$, Alice is the only player who sees two hats of the same colour. However, if $S = 1$, i.e., if all hats are of the same colour, then the two other players will also see two hats the same colour and step forward with probability p. We condition on the random variable $N = i$, for i being the number of the other players who step forward. So,

$$
\begin{aligned}
\mathrm{EU}_g[\langle 0, p, p\rangle \mid S = 1] =\\
= \sum_{i=0,1,2} \mathrm{EU}_g[\langle 0, p, p\rangle \mid S = 1, N = i]\mathrm{P}(N = i \mid S = 1). \quad (6)
\end{aligned}
$$

i	$\mathrm{EU}_g[\langle 1, p, p\rangle \mid S = 1, N = i]$	$\mathrm{P}(N = i \mid S = 1)$
0	0	$(1 - p)^2$
1	2	$2p(1 - p)$
2	2	p^2

Consequently,

$$
\mathrm{EU}_g[\langle 0, p, p\rangle] = (0)\frac{1}{2} + (0(1 - p)^2 + 2(2p(1 - p)) + 2p^2)\frac{1}{2} = -(p - 2)p. \quad (7)
$$

Now, consider

$$
\begin{aligned}
\mathrm{EU}_g[\langle 1, p, p\rangle] = \mathrm{EU}_g[\langle 1, p, p\rangle \mid S = 0]\mathrm{P}(S = 0) \\
+ \mathrm{EU}_g[\langle 1, p, p\rangle \mid S = 1](1 - \mathrm{P}(S = 0)) \quad (8)
\end{aligned}
$$

Again, $P(S = 0) = \frac{1}{2}$. Also, $\text{EU}_g[\langle 1, p, p \rangle \mid S = 0] = -2$ and $\text{EU}_g[\langle 1, p, p \rangle \mid S = 1] = 2$. So $\text{EU}_g[\langle 1, p, p \rangle] = \frac{1}{2}(2) + \frac{1}{2}(-2) = 0$ for all values of p.

The point $\langle 1, 1, 1 \rangle$ is not a Nash equilibrium, since unilateral deviation to $\langle 0, 1, 1 \rangle$ increases the group's payoff from $\text{EU}_g[\langle 1, 1, 1 \rangle] = 0$ to $\text{EU}_g[\langle 0, 1, 1 \rangle] = -(1 - 2)1 = 1$. $\langle 0, 0, 0 \rangle$ is a Nash equilibrium, since unilateral deviation leaves the group's payoff at $\text{EU}_g[\langle 0, 0, 0 \rangle] = \text{EU}_g[\langle 1, 0, 0 \rangle] = 0$. We then investigate whether there are equilibria in mixed strategies. To do so, we solve the following equation for p under the constraint $0 < p < 1$:

$$\text{EU}_g[\langle 0, p, p \rangle] = \text{EU}_g[\langle 1, p, p \rangle]$$
$$-(p - 2)p = 0. \tag{9}$$

Equation (9) has no solution under the constraint $0 < p < 1$. Hence, there exists only one symmetric equilibrium, viz. $\langle 0, 0, 0 \rangle$.

4   An Intuitive Account

So why is it that a person whose degree of belief for some proposition is $\frac{1}{2}$ should not be posting her betting rates accordingly? Why should she refrain from expressing a willingness to sell a bet for $2 that pays $4? Since we are looking for a symmetric solution, we need to consider what would happen if every player did declare herself willing to sell a bet at odds that correspond to her betting rate. This would mean that the strategy of each player would be to step forward and offer to accept the second bet if she sees two hats of the same colour. The profile of the player's strategies would thus be $\langle 1, 1, 1 \rangle$. We will now provide intuitive arguments to the effect that $\langle 1, 1, 1 \rangle$ cannot be the rational solution neither in the self-interested nor in the group-interested case.

Let us first consider the case of self-interested decision-making. There are two states — one that is favourable and one that is unfavourable for selling the bet. In the favourable state, all the hats are of the same colour, the bookie loses the bet and the player gains $2. In the unfavourable state, the hats are of different colours, the bookie wins the bet and the player loses $2. In the favourable state, three players will step forward and the bookie takes a single bet by randomising over the willing players. So there is only a $1/3$ chance that a player who steps forward will actually sell the bet. However, in the unfavourable state, only one player steps forward and she is sure to get the bet. Even though my degree of belief that the hats are of different colours is $\frac{1}{2}$, it is irrational to express my willingness to sell a bet at the matching betting rate if this willingness translates into a greater opportunity to sell the bet when I am bound to lose and a lesser opportunity to sell the bet when I am bound to win.

Consider the following analogy. Suppose that my degree of belief in the proposition that it will snow tomorrow at noon is $\frac{1}{2}$. Suppose that you are going to the bookmaker in town tomorrow morning and are willing to place a bet for me on that proposition that costs \$2 and pays \$4. I might refrain from letting you do so on grounds of the following reasoning. In the favourable state, viz. when it snows at noon, this is likely to be preceded by a cold night and there is only a small chance that your car will start in the morning. Thus you will probably be unable to drive to town and place the bet. In the unfavourable state, viz. when it does not snow at noon tomorrow, there is a very good chance that you will make it into town. Then clearly it would be irrational for me to give you the assignment to place the bet on my behalf.

In the case of group-interest decision-making, it is also irrational to express my willingness to sell a bet at the matching betting rate. Suppose that we would all do precisely that. Then I would reason as follows when seeing two hats of the same colour. In the favourable state, with all hats of the same colour, two other players will step forward and nothing is lost by my not stepping forward. In the unfavourable state, with the hats being of different colours, I am the only one who would step forward, so I can save the group from a loss by not stepping forward. Hence unilateral deviation from $\langle 1, 1, 1 \rangle$ to the pure strategy of not stepping forward improves the group payoff. We can conclude that simply stepping forward —i.e., stepping forward with probability 1— when seeing two hats of the same colour cannot be the rational strategy.

To understand why $\langle p, p, p \rangle$ is not a Nash Equilibrium for any $p > 0$, it is instructive to construct some graphs. First let us look at self-interested decision-making. We have plotted $\mathrm{EU}_a[\langle 0, p, p \rangle]$ (dashed line), $\mathrm{EU}_a[\langle 1, p, p \rangle]$ (dotted line) and $\mathrm{EU}_a[\langle p, p, p \rangle] = p\mathrm{EU}_a[\langle 1, p, p \rangle] + (1 - p)\mathrm{EU}_a[\langle 0, p, p \rangle]$ (full line) in Figure 1. Note that if the group members display even the slightest inclination to bet, then the expected value of each person's payoff is lower than if they would have refrained from betting. This is easy to understand. For the $\langle 1, 1, 1 \rangle$ strategy, we said that it was not a good idea to express one's willingness to bet if this translates in placing the bet for sure in the unfavourable situation but only having a one in three chance of placing the bet in the favourable situation. Neither is it a good idea to be inclined to express one's willingness to bet with some positive chance p, if it is the case that if one were to act on this inclination, then the expression of one's willingness would translate in placing the bet for sure in the unfavourable situation and a less than maximal chance of placing the bet in the favourable situation.

Let us now turn to the group-interested decision-making. We have plotted $\mathrm{EU}_g[\langle 0, p, p \rangle]$ (dashed line), $\mathrm{EU}_g[\langle 1, p, p \rangle]$ (dotted line) and $\mathrm{EU}_g[\langle p, p, p \rangle]$ (full line) in Figure 2. Note that for any value of $p > 0$, unilateral deviation to $\langle 0, p, p \rangle$ increases the payoff function. In this case, the issue is not about

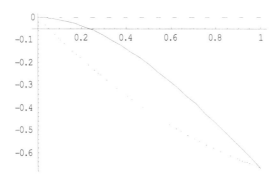

FIGURE 1. Self-Interested Decision-Making

loss avoidance. If I am determined to step forward upon seeing two hats of the same colour, then the expected payoff for the group will be zero. But the point is that I can do better on the group's behalf. If the others express their willingness to step forward when seeing two hats of the same colour or if they are inclined to do so with positive chance p, then I can exploit this by refraining from stepping forward. This would only marginally decrease the chance for a win for the group in the favourable situation when all hats are of the same colour, but it would guarantee the absence of a loss for the group in the unfavourable situation when the hats are of different colours. Since the win and the loss are equal in size and the probabilities of the favourable state is the same as that of the unfavourable state, the expected payoff to the group is positive if I decide to stay put.

5 Discussion

The general lesson is this: Willingness to bet is one thing, but a binding declaration of such willingness is another. Betting rates at which we are equally willing to buy or to sell a bet on a proposition are one thing, but posting these betting rates is a different matter. Degrees of belief might match our betting rates. (There are counter-examples to this claim as well, but nothing we say here addresses this issue.) But they certainly need not match our posted commitments to bet. The former can be expected to differ from the latter if declarations of willingness to bet do not automatically translate into betting opportunities.

In future work, we intend to explore two applications of this problem:

First, suppose the bookie 'sweetens the pie' by offering to buy the second bet at stakes $\$4-\varepsilon$ for a small ε. In this case the Nash equilibrium is $\langle p, p, p \rangle$

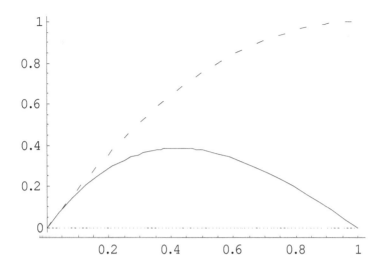

FIGURE 2. Group-Interested Decision-Making

for some p such that $1 > p > 0$ and that equilibrium is no longer the same for self-interested and group-interested decision-making: The value of p is greater for the former than for the latter. The situation has the structure of a tragedy of the commons. Self-interested fishermen tend to put too many boats on the sea and over-fish so that they can barely make a living, whereas the group-interest demands that we restrict the number of boats on the sea. Self-interested decision-makers will pick the value of p so that their expected payoff equals 0, whereas the group-interest demands lower values of p yielding a higher expected utility for the group.

Second, the structure of the decision-problem in the second bet of the 'sweetened' hats problem, with group-interested players, is similar to the decision-problem faced by juries. In the hats problem, there is a *group choice* of either selling a bet or not and we aim at selling losing bets and refraining from selling winning bets. There is an *individual choice* of either stepping forward to sell the bet or not. The *procedure* is that group choice of selling the bet is taken if and only if at least one person steps forward. The *situation* is that all hats either are single-coloured or not. Each person receives a *private signal* as to whether or not the hats might be single-coloured, but this private signal is not fully reliable. (It is not fully reliable, if both hats you observe are of the same colour.) In a jury vote, there is a *group choice* of either acquitting or convicting the defendant and we aim at acquitting the innocent and convicting the guilty. There is an *individual choice* of voting innocent or guilty. The *procedure* is that the group choice of acquitting is taken if and only if at least one person votes innocent. The *situation* is that the defendant either is guilty or innocent. Each person receives a *private signal* as to whether the defendant is guilty or innocent, but this private signal is not fully reliable. In the hats problem, the question is whether it is rational to step forward to take the bet when I receive a private signal suggesting that all hats are of the same colour — which is a private signal that the bet is favourable from the group's point of view. In the jury problem, the question is whether it is rational to step forward to vote innocent when I receive a private signal that the defendant is innocent — which is a private signal that the acquittal is favourable from the jury's point of view. There are differences in detail between the two cases, but both of them can be modelled as instances of the same formal decision-problem.

The core idea of strategic voting [1, 2, 3] is that jury members in a unanimous jury will not vote according to their private signal. Suppose they were to do so. As a jury member, my vote of innocence only matters if it is pivotal, i.e., if all others have voted guilty. But, in this case, there is overwhelming evidence that the suspect is guilty since all others must have received private signals of guilt. So even if I receive a signal of innocence, I should vote guilty. But if all reason like this, then the suspect will be convicted, even if everyone receives a signal of innocence. Still, even if one takes into consideration that the others' votes might not express their private signals, voting innocent upon receiving a signal of innocence is too rash. Just as in the 'sweetened' version of the hats problem, with group-interested players, strategic considerations force jury members to adopt a randomised strategy of voting innocent with probability higher than zero but lower than one, when they receive an innocence signal.

Our methodology to determine a randomised strategy $\langle p, p, p \rangle$ for group-interested decision-making in the 'sweetened' hats problem is (with some qualifications) the same as the methodology for determining a randomised strategy $\langle p, \ldots, p \rangle$ for voting innocent upon receiving an innocence signal in jury voting. And depending on the values of the parameters, there are indeed situations in which the rational choice in jury voting is to randomise with a surprisingly low chance of voting innocent upon receiving an innocence signal, just like the rational solution in the hats problem is to randomise with a surprisingly low chance of stepping forward to sell the bet upon seeing two hats of the same colour.

References

[1] Jeffrey S. Banks. Committee Proposals and Restrictive Rules. *Proceedings of the National Academy of Science*, 96:8295–8300, 1999.

[2] Timothy Feddersen and Wolfgang Pesendorfer. Voting Behavior and Information Aggregation in Elections with Private Information. *Econometrica*, 65:1029–1058, 1997.

[3] Timothy Feddersen and Wolfgang Pesendorfer. Convicting the Innocent: The Inferiority of Unanimous Jury Verdicts under Strategic Voting. *American Political Science Review*, 92(1):23–35, 1998.

[4] Sara Robinson. Why Mathematicians Now Care About Their Hat Color. *New York Times*, April 10 2001.

Benedikt **Löwe**, Eric **Pacuit**, Jan-Willem **Romeijn** (*eds.*)
Foundations of the Formal Sciences VI
Reasoning about Probabilities and Probabilistic Reasoning

After Dutch Books

TIMOTHY CHILDERS*

Department of Logic, Institute of Philosophy, Academy of Sciences of the Czech Republic,
Jilska 1, Prague, 110 00, Czech Republic
E-mail: childers@site.cas.cz

1 Introduction

Dutch Book arguments are sometimes purported to establish that degrees
of belief should obey the probability calculus. I will argue that they do
not, and cannot, establish this. I do this by examining several competing
interpretations of the arguments, and show that they share the same fundamental flaw, a connection to a dispositional account of belief. If we try
to remove this account, there are better types of arguments for establishing
links between degrees of belief and probabilities.

The first part of this paper covers the behaviourist interpretation of the
Dutch Book argument, while the following two cover two different types of
depragmatized Dutch Book arguments — Howson and Urbach's counterfactual interpretation and Howson's logical interpretation. While I conclude
that Howson's logical interpretation is the best that can be made of the
Dutch Book argument, in the next section I argue that there is a better argument, from de Groot, that degrees of belief should conform to the
probability calculus. This argument can be made to fit nicely with Howson's logical interpretation. However, as I note in the next section, this
argument entails significant philosophical commitments. In conclusion, I
also note that there are other arguments which can avoid these particular
philosophical commitments.

*Work on this paper was supported by grant 401/07/0904 of the Grant Agency of the
Czech Republic. I thank Ondrej Majer and an anonymous referee for helpful comments.
I also thank the organizers and participants of FotFS VI for an excellent conference.

Received by the editors: 15 October 2007; 11 April 2008.
Accepted for publication: 11 May 2008.

2   The Dutch Book argument

I take the Dutch Book argument to be composed of four elements. These are first, the bet (or a betting situation); second, an account of bets as representations of degrees of belief; third, the notion of a fair bet; and finally, the Ramsey-de Finetti theorem(s). (The following account follows the version of the argument in [11].)

A bet on a proposition is a contract between a bettor and a bookie. The bettor agrees to give the bookie something of value b, if some proposition A comes to be accepted as false, and the bookie agrees to give the bettor something of value a if the proposition comes to be accepted as true (after, perhaps, consultation with an oracle). The possible losses and gains from the bettor's perspective can be represented by a payoff table:

A	Payoff
T	$+a$
F	$-b$

Normalizing the odds $p = (b/a)/(1 + b/a) = b/(a + b)$ gives the usual form:

A	Payoff
T	$S(1 - p)$
F	$-Sp$

The number p is known as the betting quotient, S as the stake. A bet against a proposition is one with the payoff signs reversed.

Betting quotients are taken as related to degrees of belief: the longer the odds a bettor is willing to accept against a proposition, the more certain the bettor is of the truth of that proposition. A *fair* betting quotient is one the bettor believes is fair to both sides in that he or she believes the potential loss or gain to be equal for both sides of the bet. It stands to reason that if a bettor thinks a betting quotient on A is fair, he or she should be willing to take either side of the bet on A (that is, either the bet on A or against A). The Ramsey-de Finetti theorem, or the Dutch Book argument, shows that fair betting quotients must be probabilities (that one side of the bet does not lead to a sure loss only if fair betting quotients obey the Kolmogorov axioms.) A converse argument shows that if fair betting quotients are probabilities, a bettor avoids sure losses.

These four ingredients when put together, it is argued, show that the probability calculus imposes consistency constraints on our degrees of belief. (De Finetti in, e.g., [4] terms degrees of belief that obey the probability calculus *coherent*.) These constraints are claimed to give us an epistemology of great power (surveyed, e.g., in [11]).[1]

---

[1] Another way to put the argument is: degrees of beliefs are betting quotients. To

3   Degrees of belief as behaviour

I shall now concentrate on the second ingredient, the link between bets and degrees of beliefs. It has been claimed that the link if very close indeed — that bets are actually degrees of beliefs, that is, that degrees of belief are nothing other than a psychological stand-in for odds actually given on propositions. This can be fairly called a behaviourist approach, and it is well-known to be unsatisfactory, since odds actually offered my bear no relation to actual strengths of beliefs. Still, it is worth looking at the details of why the behaviourist interpretation won't work, since the reasons it won't work apply to other interpretations of the Dutch Book argument as well.

Some people do not like to bet. Consider the Reverend, who considers gambling a sin: he isn't willing to give any odds.[2] Therefore, no bet will elicit his degree of belief, since he refuses to bet. However, he does have degrees of belief. Therefore, bets and degrees of belief cannot be equated. The Reverend's unwillingness to bet is just one extreme of a more general problem.

Even if someone is willing to bet, bets will not in general add up in such a way as to allow a determination of the value of joint bets. This means that even if bets did represent degrees of beliefs of propositions, they still might not represent degrees of beliefs in combinations of those propositions. The problem stems from the non-linear value of money: combined bets denominated in money may not simply be a combination of their component bets. For example, I may prefer to have a better deal when buying multiple bets, rather than buying them singly, as is required in the Dutch Book argument for the third axiom. I may have set aside a certain amount of money aside for my month's gambling, or I may just be strongly in favour of bulk discounts.

Another way to put it is that the fair betting quotient p is a ratio derived from a particular betting situation in which the bettor is willing to

---

offer certain betting quotients is stupid. Therefore, certain degrees of beliefs are stupid. All and only betting quotients, and hence beliefs, that obey the probability calculus are non-stupid. Therefore, in order to be non-stupid, our degrees of beliefs should confirm to the probability calculus. 'Stupidity' should be understood here as a technical term meaning 'leading to a sure loss/gain'. One can deny that this technical use of 'stupidity' is a good explication of the ordinary language usage, cf. [9].

[2]And not without scriptural support, of course. Matthew (5:33-37, KJV) has Jesus disallowing emphatic speech, and so expressions of degrees of certainty, at least: "Again, ye have heard that it hath been said by them of old time, Thou shalt not forswear thyself, but shalt perform unto the Lord thine oaths: But I say unto you, Swear not at all; neither by heaven; for it is God's throne: Nor by the earth; for it is his footstool: neither by Jerusalem; for it is the city of the great King. Neither shalt thou swear by thy head, because thou canst not make one hair white or black. But let your communication be, Yea, yea; Nay, nay: for whatsoever is more than these cometh of evil." The paper [13] alerted me to this passage, Marta Vlasaková helped me find it.

risk forfeiting some amount of money b to get an amount a if the proposition being betted on is true. However, b and a will obviously not be the same in all situations. So we cannot derive a general value for p. (This problem is well known: it is perhaps most forcefully expressed in [19].)

There are three traditional responses. The first is that we should keep the amount of money involved in the bets small (I do not know the origins of this response: it can be found in [5]). But then we encounter a Goldilocks problem. If amount of money on offer is too small, I won't reveal my preferences since I can't be bothered to put out the effort to protect myself against losses (or to maximize my minimal gains). If the amount is too large we're right back where we started: I won't be willing to bet (or I may choose to hedge my bets). And this response does not address the problem of packages of bets: two separate bets might fall under threshold at which I will cease to add the value of my bets, but adding a third might push them above. The problem obviously gets worse the more bets are under consideration.

The second response is that bets should be denominated in a utility currency, in which each unit of currency has equal utility. Thus, any reasonable person would see that bets should add in the required way. Savage [18], e.g., following Smith [20], refers to the construction of such a lottery. We find some lottery mechanism which our subject is willing to say is fair, in that each of the produced outcomes, say, tickets, coloured balls or spun pointers ending up at some point on a wheel, are for him or her equally likely. These can then serve as units of currency in a bet, each unit having the value of the prize divided by the total number of outcomes.

This response makes the strong assumption that for each bettor and betting situation, a suitable mechanism for producing a utility currency can be found. But there are people like the Reverend, for whom no such lottery exists. And we still cannot rule out the odd few who collect lottery tickets, are repulsed by spinning arrows, or have other attachments or aversions to chancy devices. These people may agree to the use of the currency, but it would not reveal their degrees of belief.

The final fix for the Dutch Book argument's problems is to require that players be compelled to bet. (This seems to have been considered by de Finetti, albeit not wholeheartedly, in [5, p. 102].) But this will not get *my* true odds: having a strong aversion to betting, I will respond by giving you odds at which I think I will lose, so I don't have to accept any filthy lucre. The Reverend will likewise be unmoved, for he welcomes martyrdom. It is also difficult to see how compulsion shows that degrees of belief should follow the probability calculus: it only seems to show that, unless you are willing to be a martyr, you should try to please the compeller.[3]

---

[3]For some, this might be enough. For example, Hobbes, whose account of psychology

The reader may wonder why I am wasting time beating a dead horse — the objections, and responses, I have discussed are well known. My aim in the foregoing has not been not to repeat what has been said at length, but to clearly spell out why a link between beliefs and betting behaviour is not strong enough to justify the argument for (at least) the Dutch Book argument for the third axiom. In the next section I will argue that for the same reasons that the link is not strong enough in the literal case, it will also not be strong enough in any interpretation which takes the Dutch Book argument to be an idealization.

4   The as-if interpretation

A standard defence of the Dutch Book argument is that it portrays an ideal agent, not caught up with worries about actual money, which we should find compelling. One way to idealize the betting situation is to not require that money change hands, but only to consider what would happen were money to change hands. This particular interpretation was put forward in [11], where they gave a counterfactual interpretation of the Dutch Book argument:

> Attempts to measure the values of options in terms of utilities are traditionally the way people have sought to forge a link between belief and action, and much contemporary Bayesian literature takes this as its starting point. We do not want to deny that beliefs have behavioural consequences in appropriate conditions, they clearly do, but stating what those conditions are with any precision is a task fraught with difficulty, if not impossible... [T]he conclusion we want to derive, that beliefs infringing a certain condition are inconsistent, can be drawn merely by looking at the consequences of what *would* happen if anyone *were* to bet in the manner and in the conditions specified [11, p. 77].

This interpretation seems superior to the literal one. It appears to avoid problems involving the currencies involved in betting, as well as those associated with unwilling bettors, since no money actually changes hands. This takes us some way to removing the extraneous elements of the Dutch Book argument.[4]

---

in *Leviathan* seems to be what proponents of traditional Dutch Book arguments have in mind, famously argued that a contract entered into under compulsion is valid (in *Leviathan*, Chapter XIV). Whatever the validity of Hobbes' view, it still does not show that degrees of belief are betting odds produced under compulsion, unless we are willing to accept a psychology in which degrees of belief just are betting odds, no matter how produced.

[4]Colin Howson no longer adheres to this interpretation, as we shall see. I discuss it, however, because it seems to me that this interpretation is the best possible of its kind,

Willingness to bet in this interpretation is, of course, a counterfactual, or subjunctive, matter. The standard semantics for dealing with counterfactuals are Lewis-Stalnaker semantics. Unfortunately, using this semantics will make the Dutch Book argument for the third axiom invalid, as can easily be seen:

> If you were to bet on A you would regard p as a fair betting quotient.
>
> If you were to bet on B you would regard q as a fair betting quotient.
>
> *Therefore...*
>
> If you were to bet on A and on B, you would regard p and q as fair betting quotients.

This is an instance of the so-called counterfactual fallacy of strengthening the antecedent, that is the argument from 'If A were the case then C' to 'If A and B were the case then C'. To see that it is a fallacy, substitute "The match is struck" for A, "The match bursts into flame" for C, and "The match is soaked in water" for B. (This paragraph is based on an argument from [1]. A discussion of strengthening the antecedent of counterfactual conditionals can be found in [15, p. 17].)

The argument cannot be fixed by choosing some other semantics, as the following example shows. Consider Harold, known to all as Dirty Harry — deservedly so, given his hygienic proclivities. Dirty Harry lives in Prague, and it is summer. He is feeling suicidal (although he lives in a wonderful city, few people will approach him), and is disposed to kill himself. But, it's hot, and he is also disposed to have a beer. The beer drinking disposition may preclude his disposition for suicide. Perhaps he no longer feels suicidal because the beer is so good, or perhaps because he gets too drunk to remember his troubles. Or perhaps he accidentally stumbles in front of tram, and is run over before he can kill himself by his own hand. Conversely, suicide precludes beer drinking. So, Dirty Harry's beer drinking disposition can block his disposition to commit suicide, and vice versa.

Any logic which faithfully represents dispositions to behaviour will also represent dispositions in general not being serially or jointly realizable. This means that the Dutch Book argument, at least for the third axiom, is invalid, since it assumes that dispositions *are* so realizable. The as-if interpretation shows us exactly what is wrong with the Dutch Book argument, and why

---

and determining where it goes wrong shows why the Dutch Book argument cannot be saved. As well, this interpretation also parallels Ramsey's view that "...the meaning of a sentence is to be defined by reference to the actions to which asserting it would lead, or, more vaguely still, by its possible causes and effects." [17, p. 57].

it cannot be saved. If we represent the Dutch Book as being about consequences, then even counterfactual bets and degrees of belief won't match.
If we represent the Dutch Book argument as not being about consequences,
it's not a Dutch Book argument, since there's no contract between a bettor
and bookie.[5] If we are to save the Dutch Book argument, then, we must
remove all considerations of dispositions to bet from the argument, and so
of money, no matter how abstractly considered.

5   The 'logical' interpretation

Ramsey had made the claim that subjective probability involved consistency, like deductive logic. Logic has already been de-psychologized, so
taking a cue from Ramsey, we might find some inspiration from logic for
the interpretation of the Dutch Book argument. Colin Howson has recently
taken up this task (e.g., [10, 11]).

First, a familiar fact: a truth valuation for a set of sentences is consistent
if it can be extended to cover all sentences in the language in accordance with
the basic semantic definition, which lays down rules for truth assignments.
To use Howson's example, if we were, using classical logic, to evaluate $A \rightarrow B$ and A as true, but B as false, there would be no assignment of truth
values to all the other sentences of the language in accordance with the
basic semantic definition. As he puts it, the problem is to solve a system of
equations where we are given $v(A \rightarrow B) = 1$, $v(A) = 1$ but $v(B) = 0$. There
is no such solution to such a system of equations of truth assignments, and
so it is inconsistent.

According to Howson, the parallel notion of valuation for probability is
that of assignments of fair betting quotients to propositions. An assignment
of betting quotients to a set of propositions is consistent if it can be extended
to an assignment over all propositions. Betting quotients serve as a model
for degrees of belief in the same way that truth values serve to model the
notion of truth: they share some common features of interest. But just as
a truth valuation is mostly independent of a particular theory of truth, so
a valuation of fair betting quotients is intended to be free of a substantial
theory of uncertainty. Betting quotients are meant to serve as the semantic
correlates of degrees of beliefs, not by any particular assignment to the
betting quotients, but as groups of possible distributions of fair betting
quotients. The notion of sure loss is no longer tied to eliciting particular

---

[5] If the Dutch Book is a dramatization, then either it has drama or not. With drama
it is invalid. Without drama it is pointless. I am only claiming that an idealization of the
Dutch Book argument that removes the notion of consequences is fatal to the argument.
There are other idealizations of the argument that are, for some purposes, reasonable.
For example, we could assume the bettor is logically omniscient, has betting quotients
defined over all the states of the world, can tell the difference between the odds 89762:1
and 89761:1, and has access to an oracle to settle wagers.

degrees of belief, but serves as a heuristic to interpret fairness, as truth in a valuation is not tied to any particular distribution of truth values.

Given a particular assignment of fair betting quotients, we find that it can only be extended to all propositions if it is fair. But we also know that betting quotients can only be fair if (and only if) they obey the rules of the probability calculus. Howson draws from this the lesson that we can treat fair betting quotients as semantic objects, the analogue of truth values. The associated syntax is simply the probability calculus. The Ramsey-de Finetti theorems serve as a kind of soundness and completeness theorems, showing the syntax and the semantics to be in complete agreement.

Adherence of degrees of belief to the probability calculus is therefore consistency, and this adherence is to be justified in the way that adherence to logical consistency is. According to Howson, after Frege adherence to consistency is not justified by appeal to actual or imagined consequences. Instead, it is justified in terms of adequately capturing some feature of reasoning of interest. We can amplify on Howson's account. There is no angry god of logic who hurls lightening bolts at the inconsistent. There will always be circumstances where inconsistency is harmless (or perhaps even helpful). Therefore, attempts to justify logic in terms of (good or bad, imagined or real) consequences will always fail. Instead, logic is a tool for exploring or modelling certain objects of interest, like, e.g., certain notions of truth. Similarly, attempts to link probability and consequences are doomed. But probability is useful for modelling certain features of uncertainty.

This is a non-psychological, depragmatized, reading of the Bayesian interpretation: it is about the assignment of numbers, called fair betting quotients, to propositions. These fair betting quotients may serve as a heuristic, or as an explication of certain aspects of uncertainty. But, like the notion of truth in logic, the notion of uncertainty in Bayesianism remains, according to Howson, to a large degree independent of a substantive theory of uncertainty.

This interpretation of the Dutch Book argument of course severs any link between behaviour and probability on the one hand and belief on the other. It does so because belief does not, in fact, figure in the argument. The link between fair betting quotients and beliefs must therefore be provided by another argument. But we have already seen that the usual way of making such a link won't work.

We are interested in logics because they serve to model certain notions we have, e.g., about truth. In fact, the link between the basic semantic definitions of first order logic and certain notions of truth is obvious, even if it is not the last word (which is why we are interested in different types of logics). In the case of fair betting quotients and degrees of belief we have a link provided by behaviour, but not by anything else. In other words, it is hard to see what appeal fair betting quotients have as semantic objects, other than their being related to degrees of belief by betting, actual or not.[6]

6   Probability from likelihood

Dutch Book arguments (and, for similar reasons, utility theoretic arguments) will not provide the requisite link between probabilities and beliefs. There are, however, alternatives if we consider different conceptions of belief. For example, some take belief to be a representation — a propositional attitude, as opposed to a disposition. Individual beliefs and sets of beliefs are structured in certain ways. One part of this structure is that belief comes in degrees: we believe some things are more likely than others. Using a likelihood ordering and a means of calibration, Morris de Groot [6] showed how to build a representation theorem for degrees of belief (which has mostly been ignored, excepting [7] and [8]).

The basis of the representation theorem is an ordering of beliefs, in terms of a likeliness relation, where the likeliness relation obeys the axioms of qualitative probability, and the propositions of the beliefs form a field. (It also requires that the certain event is more likely than the impossible event, and that every event is as least as likely as the impossible event.) The next step is to calibrate degrees of uncertainty by putting the members of the algebra into correspondence with a reference algebra with a known probability distribution. One way of doing this, as French showed, is to use a visual representation. Although nothing turns on which representation we use, French uses a wheel of fortune — a disk with a (perfectly balanced and oiled) spinning arrow. The arrow's landing at some particular point is an event. We can then construct a field of events which contains points, intervals and combinations of intervals. Using the obvious flat distribution over the events, we can use the normalized length of the arcs to calibrate the uncertainty in the original algebra by matching up each event with a point or an interval on the wheel, equating the certain event with the entire circumference of the wheel, and the impossible event with length 0. Probability then turns out to be the percentage an event takes up on the circumference of the wheel of fortune.

Just as the Dutch Book argument is tied to an account of disposition, the de Groot construction is tied to an account of beliefs as attitudes towards proposition-like entities. The constraints of the construction are taken as

---

[6]I do not believe that there is a strong distinction between 'pragmatic' and 'epistemic' interpretations of the Dutch Book argument, since any epistemic theory without consequences is hardly of interest. But theories of truth without discernable pragmatic consequences are of interest. This is why truth in a model is more closely related to the notion of truth than a depragmatized notion of betting is linking to the notion of belief. Of course, truth may be pragmatic notion. But it would be a severe theory that grounded truth on immediate consequences, which is what the Dutch Book does with belief.

plausibly being an explication of the structure of the attitudes and their contents. This fits well with a logical interpretation like Howson's. The underlying semantics are proposition-like entities with some structure. The linking of the events in the underlying algebra with the reference distribution serves to show that the probability calculus completely captures the underlying likeliness relation. There is no longer any need for actualizations of dispositions, and the semantics of the likeliness relation is naturally linked to the syntax of the probability calculus.

I am not arguing that we now have set once and forever the structure of partial beliefs. Two obviously questionable assumptions are the total ordering of beliefs and the sharp correlation of the reference experiment and the algebra of beliefs. Further investigation is required to show in which ways the construction can be plausibly weakened and generalized.

7   Problems with probabilities from likelihood, further prospects

There are several reasons why one might express scepticism about the prospects of this representationalist programme of justification. The representationalist account of belief entails significant philosophical commitments. It takes beliefs to be open to introspection. It further assumes that beliefs are proposition-like entities, or at least that they can be represented as such. It requires the assumption of a flat distribution. Since there is no element of risk or compulsion, one could lie about one's degrees of beliefs.

I begin with the last: how, absent a sanction, we can give a reason for someone to conform their degrees of beliefs to the probability calculus? The traditional Dutch Book argument at least aims to show that if your degrees of belief are somehow not aligned with the probability calculus, something bad will happen to you. There are no sanctions in the de Groot construction.[7] So according to the representationalist account, someone could lie about their degrees of belief, or even have incoherent beliefs (which they do not reveal). However, if I am right, the Dutch Book argument also does not establish that someone could not lie about their beliefs, much less suffer from those lies. It is certainly true that we cannot determine if someone accurately represents their degrees of beliefs, but this is beside the point, for representationalist does not equate belief with action.

The question of introspection is much trickier, turning as it does on a major debate in the philosophy of mind. The representationalist view does indeed entail significant philosophical commitments, as does a dispositional

---

[7]I should also point out that, even though I do not advocate such an approach, sanctions can be added to the construction by introducing penalties via the reference experiment. For example, using the wheel of fortune, areas can be normalized to betting quotients, and the usual machinery then applied.

view of belief. The point of this paper is that views about what beliefs are determine what sorts of arguments should be made for equating degrees of belief with probabilities. It is true that it would be a strong assumption that we can always conjure up an appropriate representation to scale the probabilities. Even though flat distributions, which can be used as reference distributions, seem very natural in many cases (witness the popularity of logical interpretations of probability based on the maximum entropy principle), they need not apply in every case. Indeed, it probably is the case that there will be no appropriate reference distribution for some propositions: but then it is a strong assumption that there should be a corresponding probability for all propositions.

I have argued that the search for universal sanctions to enforce the probability calculus is pointless. Instead we might undertake the task of explicating the notion of partial belief relative to a given set of purposes. This is, in fact, what most accounts of interpretations of probability focus upon, at least in the philosophy of science. So instead of wondering which way we may be struck down for heresy, we question whether the Bayesian solution to the Duhem-Quine is the correct one. In other words, the foregoing objections turn on what one expects from an interpretation of probability. If we take it that an interpretation should tell us when, on pain of some penalty, we should obey the probability calculus, then a de Groot style representation is probably not ideal. If, however, we take interpretations of probability to be explications of certain notions about belief, then a de Groot style representation might be what we want given a representationalist account.

Still, those unhappy with the Dutch Book argument, and those uncomfortable with propositional attitudes might wish to simply drop beliefs out of the picture completely, and concentrate on something else. There are several alternatives, the most prominent being James Joyce's [12] and Cox's [2]. Neither of their constructions refers directly to belief, and so are free of the difficulties inherent its explication. If we approach these arguments as not being about belief, then presumably they will be taken as explications of pre-theoretical notions. The notion of there being such explications is very much out of fashion. But if we are to make any headway in debates over the foundations of subjective probabilities, then the debate must take place at the level of the explication of belief, or one of its substitutes.

References

[1] Paul Anand. *Foundations of Rational Choice under Risk*. Clarendon Press, 1993.

[2] Richard T. Cox. *The Algebra of Probable Inference*. Johns Hopkins University Press, 1961.

[3] David Mellor, editor. *Frank Plumpton Ramsey's Foundations: Essays in Philosophy, Logic, Mathematics and Economics*. International Library of Psychology, Philosophy and Scientific Method. Routledge & K. Paul, 1978.

[4] Bruno de Finetti. Sul Significato Soggettivo della Probabilità. *Fundamenta Mathematicae*, 17:298–329, 1931. Translated as "On the subjective meaning of probability", in: [16, pp. 291-321].

[5] Bruno de Finetti. La prévision: ses lois logiques, ses sources subjectives. *Annales de l'Institut Henri Poincaré*, 7:1–68, 1937. Translated as "Foresight: Its Logical Laws, Its Subjective Sources" in: [14, pp. 93–158].

[6] Morris H. DeGroot. *Optimal Statistical Decisions*. McGraw-Hill, 1970.

[7] Simon French. On the Axiomatisation of Subjective Probabilities. *Theory and Decision*, 14(1):19–33, 1982.

[8] Simon French. *Decision Theory: An Introduction to the Mathematics of Rationality*. Ellis Horwood, 1988.

[9] Alan Hájek. Scotching Dutch Books? *Philosophical Perspectives*, 19(1):139–151, 2005.

[10] Colin Howson. Probability and Logic. *Journal of Applied Logic*, 1(3–4):151–165, 2003.

[11] Colin Howson and Peter Urbach. *Scientific Reasoning: The Bayesian Approach*. Open Court, 3rd edition, 2005.

[12] James Joyce. A Nonpragmatic Vindication of Probabilism. *Philosophy of Science*, 65(4):575–603, 1998.

[13] Ivan Kramosil. Some Remarks on Philosophical Aspects of Laws of Large Numbers. In Timothy Childers, Vladimír Svoboda, and Petr Kolář, editors, *Logica '95: Proceedings of the 9th International Symposium*, pages 179–197, 1996.

[14] Henry Kyburg and Howard Smokler, editors. *Studies in Subjective Probability*. Wiley, 1964.

[15] David Lewis. Counterfactuals and Comparative Possibility. *Journal of Philosophical Logic*, 2:418–446, 1973.

[16] Paola Monari and Daniela Cocchi, editors. *Bruno de Finetti. Probabilità e Induzione*. Editrice CLUEB, 1993. Suppl. 3 of volume 52 (1992) of the journal *Statistica*.

[17] Frank P. Ramsey. Facts and Propositions. *Aristotelian Society Supplementary*, VII:153–157, 1927. reprinted in: [3, pp. 40–57].

[18] Leonard J. Savage. Elicitation of Personal Probabilities and Expectations. *Journal of the American Statistical Association*, 66:783–801, 1971.

[19] Frederic Schick. Dutch Bookies and Money Pumps. *Journal of Philosophy*, 83:112–119, 1986.

[20] Cedric A.B. Smith. Consistency in Statistical Inference and Decision. *Journal of the Royal Statistical Society*, 23:1–37, 1961.

Benedikt **Löwe**, Eric **Pacuit**, Jan-Willem **Romeijn** (*eds.*)
Foundations of the Formal Sciences VI
Reasoning about Probabilities and Probabilistic Reasoning

Probabilistic Epistemic Chains

JEANNE PEIJNENBURG

Faculteit Wijsbegeerte, Rijksuniversiteit Groningen, Oude Boteringestraat 52, 9712 GL Groningen, The Netherlands
E-mail: jeanne.peijnenburg@rug.nl

Within epistemology it is customary to distinguish between conditional and unconditional beliefs or propositions. Conditional propositions are inferential: they are justified on the basis of other propositions. Unconditional beliefs are noninferential: they are the notorious basic beliefs that are part and parcel of any epistemology of the foundationalist type. They may be known a priori, or incorporate sensory experience, or be 'self-justifying' in one way or another — the important thing is that they serve as the foundation on which the edifice of our knowledge can be erected.

Condider the simplest of these edifices, the linear epistemic chain: proposition S_0 is justified by S_1, which in turn is justified by S_2, and so on. Then, according to foundationalists, there must be some proposition S_{n+1} that lies at the base of the chain and that is itself unconditionally justified. Denying the necessity of such a basic proposition would amount to accepting an infinite regress, and the ban on that is a principal rule in the history of epistemology. In fact, an aversion to infinite regresses is the raison d'être of foundationalist theories from Aristotle and Descartes to Locke, Kant, Russell and C. I. Lewis.[1]

C. I. Lewis is no doubt the most famous and influential foundationalist of the twentieth century. He repeatedly argued with fervour that an epistemic chain must end in a proposition that is certain or has probability unity: $P(S_{n+1}) = 1$. (Like Lewis, I do not distinguish between propositions 'being certain' and 'having probability one'.) If the chain were not to be so terminated, Lewis argued, then the probability of the proposition at the other end of the chain, S_0, would be zero [12, p. 172]. Because of his insistence on the grounding of epistemic chains in certainties, Lewis is sometimes called

---

[1] In this paper I will neglect the classic rival of foundationalism, namely coherentism, the ancestry of which is no less impressive.

Received by the editors: 29 July 2007; 26 December 2007.
Accepted for publication: 1 January 2008.

a *strong* foundationalist (by [1, pp. 26–27]) or a *classic* foundationalist (by [2, pp. 53ff]).

Contemporary foundationalists, however, are mostly *moderate* foundationalists [1, p. 26]. Like their philosophical forbears, they believe that epistemic chains only make sense if they come to an end in a basic proposition, but unlike them they hold that this basic proposition need not have probability one. It is enough if it has a definite probability greater than zero, $P(S_{n+1}) > 0$, and if it is such that it probabilistically supports its neighbour proposition S_n, so that $P(S_n|S_{n+1}) > P(S_n|\neg S_{n+1})$.[2] But although moderate foundationalists are dealing with probabilistic rather than non-probabilistic epistemic chains, their abhorrence of infinite chains is every bit as great as that of the strong foundationalists. If a proposition S_0 is probabilistically justified by S_1, which in turn is probabilistically justified by S_2, and so on, then there must be a proposition S_{n+1} such that $P(S_{n+1}) > 0$, and $P(S_n|S_{n+1}) > P(S_n|\neg S_{n+1})$, on pain of making nonsense of the entire chain.

In several papers Peter Klein has argued against what he calls the "infiniphobia" of old and new foundationalists [9, p. 728]. He champions 'infinitism', a position which "has never been advocated by anyone with the possible exception of Peirce" ([8, p. 306]; cf. [7, 10, 11]). Contrary to all but everyone else in epistemology, infinitists à la Klein do not fear endless epistemic chains since they do not require that such chains need be completed. What is more, they believe that actual completion would imply a rejection of infinitism [7, p. 920]. In their view the structure of justificatory reasons is "infinite and non-repeating" [8, p. 298], by which they oppose not only the foundationalist opinion that the chain must come to an end, but also the coherentist stance that at some point it must eat its own tail. Epistemic justification, whether probabilistic or not, is for the infinitist essentially conditional: claiming that a particular proposition S_{n-1} is (partially or probabilistically) supported means no more than being able to point to another proposition S_n that bestows the support in question, and that in turn receives support from still another proposition. Unconditional propositions are altogether impermissible: "There are no ultimate, foundational reasons. *Every* reason stands in need of another reason" [7, p. 299].

In recent literature there is no dearth of objections against infinitism. In [8], Klein discusses four of those objections, "beginning with the least troubling and moving to the most troubling" [8, p. 306]. As it turns out, the most troubling objection is the 'Specter of Skepticism Objection'. It states that "a proposition [is] justified only when it results from a process of justification that has been concluded" [8, p. 314]; and since infinite chains

---

[2] *Weak* foundationalism, which is the third kind of foundationalism that Bonjour distinguishes, does not satisfy this condition for probabilistic epistemic support.

of reasons cannot be concluded, no proposition could ever be justified. Klein cites his foundationalist opponent Richard Fumerton, who gave a particularly challenging wording of this objection:

> ... finite minds cannot complete an infinitely long chain of reasoning and so, if all justification were inferential, no-one would be justified in believing anything at all to any extent whatsoever. [3, p. 57][3]

Klein comments:

> "This objection to infinitism implicitly appeals to a principle that we can call the *Completion Requirement*: In order for a belief to be justified for someone, that person must have actually completed the chain of reasoning that terminates the belief in question. The infinitist cannot accept the Completion Requirement because it is clearly incompatible with infinitism." [8, p. 314].

However, contrary to what Fumerton and Klein believe, there are many infinite probabilistic chains that can be completed.[4] In the rest of this paper, I will describe one of them. Since both foundationalists and infinitists hold that infinite probabilistic chains cannot be completed, I consider my example as a counterexample to foundationalism and infinitism alike.

Suppose, then, that there is an infinite chain of propositions S_n, with $n = 0, 1, 2, \ldots$ etc. Suppose it is not known whether any of them are true or false, or even whether any of them are *probably* true or false, i.e. the unconditional probability $P(S_n)$ is unknown for all n. However, it is assumed that the conditional probabilities α_n and β_n

$$\alpha_n = P(S_n|S_{n+1})$$
$$\beta_n = P(S_n|\neg S_{n+1})$$

are known for all n. We may write

$$P(S_n) = \alpha_n P(S_{n+1}) + \beta_n P(\neg S_{n+1}), \qquad (1)$$

where, as said, we do not know the unconditional probabilities $P(S_{n+1})$ and $P(\neg S_{n+1})$. If we now write $P(\neg S_{n+1})$ as $1 - P(S_{n+1})$ and introduce the abbreviation

$$\gamma_n = \alpha_n - \beta_n = P(S_n|S_{n+1}) - P(S_n|\neg S_{n+1}),$$

---

[3]Cited by [8, p. 314]; the same quotation can be found in [5, p. 40]; [6, p. 2]; and [4, p. 150].

[4]This means that infinitism is not restricted to "thin" justification, as Klein thinks, but that it is perfectly compatible with "relatively thick" and even with "extremely thick" justification, the latter being the view which Klein deems incorrect. See [8, p. 322].

then Equation (1) becomes:

$$P(S_n) = \beta_n + \gamma_n P(S_{n+1}). \tag{2}$$

In order to calculate $P(S_0)$, one can simply iterate (2), obtaining

$$P(S_0) = \beta_0 + \gamma_0\beta_1 + \gamma_0\gamma_1\beta_2 + \ldots + \gamma_0\gamma_1 \ldots \gamma_{n-1}\beta_n + \gamma_0\gamma_1 \ldots \gamma_n P(S_{n+1}). \tag{3}$$

If $P(S_{n+1})$ were known, (3) could be used to calculate $P(S_0)$. But the difficulty is of course that $P(S_{n+1})$ is not known. In contradistinction to the first $n+1$ terms of (3), which are all conditional probabilities and therefore known, the last term in (3), like the last term in (2), contains the unknown unconditional probability $P(S_{n+1})$. So it would seem that we have come to a dead end, and that strong and moderate foundationalists are right after all. The former claim that the iteration can only be terminated by a supposed certainty, $P(S_{n+1}) = 1$; the latter say that a smaller value will serve, e.g. $P(S_{n+1}) = \frac{1}{4}$, but both insist that a definite value of $P(S_{n+1})$ is necessary to calculate $P(S_0)$.

However, there are many counterexamples that demonstrate the falsity of these foundationalist claims. Elsewhere I have given a counterexample based on an exponential series, and also one using a geometrical series [13]. However, in the geometrical example (although not in the exponential one) the conditional probabilities are uniform, i.e. they are the same from link to link of the probabilistic chain. The following example is neither geometrical nor exponential. In addition, it also avoids the artificial restriction of uniformity, for neither α_n nor β_n is uniform: α_n as well as β_n varies with n, which clearly reflects the more realistic situation.

Take

$$\alpha_n = 1 - \frac{1}{n+2} + \frac{1}{n+3} \qquad\qquad \beta_n = \frac{1}{n+3} \tag{4}$$

which is interesting, for it corresponds to a situation in which $P(S_n|S_{n+1})$ tends to 1, and $P(S_n|\neg S_{n+1})$ tends to 0 as n tends to infinity. (These features are not shared by the two earlier examples.)

In order to calculate (3), we may first evaluate its last two terms. The remainder term of (3) is the unknown probability $\gamma_0\gamma_1 \ldots \gamma_n P(S_{n+1})$. However, the coefficient of this term can be computed as

$$\gamma_0\gamma_1 \ldots \gamma_{n-1}\gamma_n = \frac{1}{n+1}. \tag{5}$$

The penultimate term of (3) can be written as

$$\gamma_0\gamma_1 \ldots \gamma_{n-1}\beta_n = \frac{1}{2}\left[\frac{1}{n+1} - \frac{1}{n+3}\right], \tag{6}$$

and from this the other terms in (3) can be obtained by substitution of the relevant values of n. This yields

$$\begin{aligned}
P(S_0) &= \frac{1}{3} + \frac{1}{2}\left[\frac{1}{2} - \frac{1}{4} + \frac{1}{3} - \frac{1}{5} + \ldots \frac{1}{n+1} - \frac{1}{n+3}\right] + \frac{1}{n+1}P(S_{n+1}) \\
&= \frac{1}{2}\left[\frac{2}{3} + \frac{1}{2} + \frac{1}{3} - \frac{1}{n+2} - \frac{1}{n+3}\right] + \frac{1}{n+1}P(S_{n+1}). \tag{7}
\end{aligned}$$

As n becomes greater and greater without bound, so $\frac{1}{n+1}$, $\frac{1}{n+2}$ and $\frac{1}{n+3}$ become smaller and smaller. Allowing n to tend to infinity in (7), we are left with

$$P(S_0) = \tfrac{1}{2}[\tfrac{2}{3} + \tfrac{1}{2} + \tfrac{1}{3}] = \tfrac{3}{4}.$$

So the value of $P(S_0)$ can be exactly calculated, even though $P(S_0)$ is determined by an infinite chain of conditional probabilities only. This gives the lie to all the claims that an infinite regress of conditional probabilities does not make sense (strong and moderate foundationalists), or always leads to zero (Lewis), or cannot be completed (Klein-style infinitists).

References

[1] Laurence Bonjour. *The Structure of Empirical Knowledge.* Harvard University Press, 1985.

[2] Jonathan Dancy. *An Introduction to Contemporary Epistemology.* Blackwell, 1985.

[3] Richard Fumerton. *Metaepistemology and Skepticism.* Rowman & Littlefield, 1995.

[4] Richard Fumerton. Epistemic Probability. *Philosophical Issues,* 14:149–164, 2004.

[5] Richard Fumerton. *Epistemology.* Blackwell, 2006.

[6] Richard Fumerton. Foundationalist Theories of Epistemic Justification. In Edward N. Zalta, editor, *Stanford Encyclopedia of Philosophy,* Edition Fall 2008.

[7] Peter Klein. Foundationalism and the Infinite Regress of Reasons. *Philosophy and Phenomenological Research,* 58:919–925, 1998.

[8] Peter Klein. Human Knowledge and the Infinite Regress of Reasons. *Philosophical Perspectives,* 13:297–325, 1999.

[9] Peter Klein. When Infinite Regresses are *not* Vicious. *Philosophy and Phenomenological Research,* 66:718–729, 2003.

[10] Peter Klein. What is Wrong with Foundationalism is that it Cannot Solve the Regress Problem. *Philosophy and Phenomenological Research,* 68:166–171, 2004.

[11] Peter Klein. Infinitism Is the Solution to the Regress Problem. In Matthias Steup and Ernest Sosa, editors, *Contemporary Debates in Epistemology*, pages 131–140. Blackwell, 2005.

[12] Clarence Irwin Lewis. The Given Element in Empirical Knowledge. *Philosophical Review*, 61:168–172, 1952.

[13] Jeanne Peijnenburg. Infinitism Regained. *Mind*, 116:597–602, 2007.

Benedikt **Löwe**, Eric **Pacuit**, Jan-Willem **Romeijn** (*eds.*)
Foundations of the Formal Sciences VI
Reasoning about Probabilities and Probabilistic Reasoning

The Role of Intuitive Probability in Scientific Theory Formation

HELEN DE CRUZ

Hoger Instituut voor Wijsbegeerte, Centrum voor Logica en Analytische Wijsbegeerte,
Dekenstraat 2, Bus 03220, 3000 Leuven, Belgium
E-mail: helen.decruz@hiw.kuleuven.be

1 Introduction

As early as the 1950s, the economist Herbert Simon [30] argued that human behaviour cannot adequately be described as 'rational', especially when rationality is defined by the basic axioms of utility theory. Contrary to what these axioms predict, humans do not always choose the best action in accordance with their desire, and sometimes even violate their own preferences. Instead, Simon [30] proposed that it may be more productive to think of rationality as *bounded*: as the world is too complex to be understood in its entirety by organisms limited in space, time and cognitive resources, they often resort to heuristics to make the world more tractable. This usually enables them to act adaptively under uncertainty, but it sometimes leads to characteristic biases.

Since the 1970s, a large number of empirical studies on heuristics and bounded rationality have focused on intuitive probabilistic judgments. As we shall see in the next section, these studies reveal a discrepancy between intuitive and formal modes of probabilistic reasoning: untutored individuals exhibit systematic deviations from formal probability theory. The cognitive science literature has hitherto mainly focused on the 'negative' aspects of intuitive probabilistic reasoning [33], in particular its inability to predict the likelihood of single events. However, in recent years, some authors (e.g., [6, 13]) have defended the view that intuitive probabilistic reasoning may also be efficient: it helps people to make adaptive decisions when faced with uncertainty. The aim of this paper is to examine whether intuitive probabilistics also play a role in scientific practice. I will defend the view that

Received by the editors: 2 September 2007; 10 February 2008, 23 February 2008.
Accepted for publication: 15 March 2008.

intuitive probabilistic reasoning plays a legitimate role in guiding scientific creativity, as it can help to develop novel ideas in the absence of adequate mathematical models in which such ideas can be phrased more precisely.

I will begin with a brief overview of the experimental literature on intuitive probabilistic reasoning, focusing on two contrasting views on its heuristic value: on the one hand, cognitive scientists such as Kahneman and Tversky [18] emphasize how intuitive probabilistic reasoning can lead to nonnormative judgments, whereas evolutionary psychologists, such as Gigerenzer and Hoffrage [12] hold that natural selection has equipped us with efficient heuristics to act adaptively under uncertainty. I next take two historical examples from scientific practice, Darwin's theory of natural selection [9] and Zahavi's handicap principle [37] to examine the role of intuitive probabilistic judgments. I show that these judgments constitute an important part of their scientific creativity.

2 Intuitive probabilistic reasoning

Making decisions in our everyday life involves generating and using probability estimates. Are the odds of getting this job high, low or somewhere in between? Given a 30% chance of rain in the weekend, should we go on with our planned camping trip? If I send a reminder to the editor of a journal which has my article under review for over four months, do I actually lower my chances of getting accepted? A large body of research, comprising hundreds of peer-reviewed publications, indicates that naive subjects tend to make nonnormative probability estimates. Note that the term 'naive subjects' refers to people who have not acquired an explicit mastery of the probability calculus; it does not in any way refer to a deficit in their cognitive capacities. The scientists I will discuss do not have this mastery and are thus by definition naive. Nonnormative probabilistic judgments are those judgments that deviate from what we deem normative according to various models of decision making and formal probability. Although constraints on working memory and processing mishaps are partly responsible for these mistakes, their systematic nature suggests that there may be more systematic biases at work. Indeed, if they were entirely due to processing limitations, the errors should be randomly distributed; instead people systematically make the same kinds of mistakes regardless of individual differences in intelligence [32]. In what follows, I discuss some examples of intuitive probabilistic reasoning, many of which are drawn from Tversky and Kahneman's seminal 1974 paper [33], which sparked much of the subsequent research on intuitive probabilistics.

Judgments by representatives Suppose you hear a description of John which goes as follows "John is a shy, withdrawn and helpful man with little interest in the world or reality. A meek and tidy person, he has a maniacal passion for detail and a need for order and structure." Is John a farmer or a librarian? Most people assess the probability that John is a librarian far greater than that he is a farmer, even though there are more farmers than librarians in a typical western population. Their conclusion is drawn from a stereotypic image of farmers and librarians. An expert in probability calculus would likely refuse to answer this question because there is no detailed information on base rates, i.e., the precise rates of farmers and librarians in the population.

Neglect of base rates Naive subjects often neglect base rates altogether. Tversky and Kahneman [33] gave subjects descriptions of five personalities, drawn from a sample consisting of 70 engineers and 30 lawyers, and asked whether the descriptions were more likely to fit engineers or lawyers. Sometimes the description did not conform to either stereotype, as in, "Dick is married, 30 years old, and has no children. He is well liked by his colleagues, highly motivated and excellent at his job". In these cases, people simply judge the odds that Dick is an engineer at 50%, entirely ignoring the base rates of the frequencies of these professions in the sample.

Misconception of chance People have the intuitive belief that randomness is a self-correcting process, in which deviations in one direction will always be corrected by deviations in the other. A couple who have two daughters often erroneously think that it is more probable that their next child will be a son.

Conjunction fallacy A fundamental principle of probability is the inclusion rule: if A includes B then the probability of B cannot exceed the probability of A, namely $P(A \& B) \leq P(A)$, because A & B is a subset of A. Subjects violate this rule in the well-known Linda problem: imagine Linda, a young woman, who is single, outspoken and deeply concerned with issues of discrimination and social justice. They are asked which of the following two hypotheses is more likely:

$$\text{Linda is a bankteller} \tag{2.1}$$

$$\text{Linda is a feminist bankteller} \tag{2.2}$$

Since the conjunction bankteller AND feminist is a subset of bankteller, hypothesis (2.1) is more likely, yet most naive subjects prefer hypothesis (2.2).

Vividness of detail bias In formal probability theory, information that does not affect the probability outcome is not relevant for probabilistic judgment. Yet, many real-world situations show that people do not take potentially beneficial risks when information is lacking: excessive brand loyalty (even when the cost of trying another potential better brand is relatively low) is a typical example, as is the 'home country investment bias', the tendency to invest too much in one's own country, instead of in internationally diversified funds [5]. In accordance with this, people tend to accord higher probability to accounts that have a vividness of detail. Imagine you receive two conflicting accounts of a particular brand of fridge. The first is in very general abstract terms, stating that the overwhelming majority of people who bought this brand were happy, and that the customer satisfaction compares favourable to other brands of the same price range. The other, however, is a heartfelt and emotional testimony of a friend who is unhappy with her fridge. In this case, there is a significant chance that you will refrain from buying it. Interestingly, Wakslak et al. [34] found that the reverse is also true: people tend to represent unlikely events in more abstract terms than events that are likely.

Although people with training in statistics or mathematics are less liable to make nonnormative probability estimates, research shows that their training does not eliminate intuitive biases. Shimojo and Ichiwaka [29] tested undergraduates with a basic training in statistics with very difficult problems, and found that they too showed biases. Just like visual illusions are difficult to resist, our naive probability intuitions seem difficult to eradicate. There is an interesting parallel with the work on naive physics by McCloskey et al. [22]. They found that students with an undergraduate training in physics made erroneous predictions about physical events that were in accordance with Aristotelian rather than Newtonian principles (e.g., the idea that a ball, released from a sling would continue to fly in a curvilinear path), as did people who had not received any formal training in physics.

3 The biased human and the rational bumblebee: are humans bad intuitive statisticians?

What are we to make of these systematic deviations of probability calculus? The seemingly straightforward conclusion, defended by Kahneman and Tversky [18], is that humans suffer from a persistent cognitive illusion: the way our brains process probabilities systematically leads to errors and biases. If true, this should have considerable implications for the human rationality debate, because it implies that humans systematically behave irrationally when probabilities are concerned. However, two alternative explanations can account for the poor performance in these tests. One possibility is that shallow heuristics may be more efficient than probability calculus in everyday situations. Johnson-Laird et al. [17] found that naive individuals are better at representing what is true than what is false. Given that true facts are usually more relevant than false facts, this heuristic is less cognitively costly than representing all possible states. Thus, we can explain the heuristics as optimal, given constraints in the time and energy organisms can devote to solving probabilistic problems. Evolutionary psychologists such as Cosmides and Tooby [6] and Gigerenzer and Hoffrage [13] go one step further, arguing that natural selection has provided us with a mental mechanism that derives correct statistical inferences. Detecting statistical regularities could have a significant impact on an organism's fitness. For example, rufous hummingbirds (*Selasphorus rufus*) need to remember how long it takes for a flower to replenish itself with nectar as a function of how many times it was emptied. In this case, Cosmides and Tooby [6] argue, there will be strong selective pressure on nervous systems to detect statistical correlations within the species' environment. Indeed, rufous hummingbirds are able to make statistical correlations between flower types, nectar concentration and replenishing time; they optimize their foraging behaviour accordingly [3].

A considerable body of experimental evidence from animal studies shows that organisms such as ants and ducks behave as good intuitive statisticians, sensitive to changes in frequency distribution in their environment, such as the occurrence of food. This is even the case for animals with a relatively simple nervous system. In his classical study of bumblebee (*Bombus spp.*) foraging, Real [26] found that the bumblebee's brain is designed to efficiently exploit floral resources from spatially structured environments. Bumblebees, like honeybees, live in colonies, with sterile workers devoted to collecting nectar and pollen, but unlike bees, they do not communicate with each other about resources. When presented with artificial flowers of different colours filled with nectar, they prefer those flowers that have a higher probability of containing nectar, even if the overall yield of both floral types is identical, i.e., bumblebees employ a risk-aversive strategy. When both types yield equal rewards, they visit them in equal proportion. Interestingly, however, when one type of flower is more common, they tend to neglect the other type even though both produce equal rewards. This may seem puzzling to us, but it makes sense from the bumblebees' point of view, because of their limitations in working memory: when both types yield the same reward, concentrating on the common type is more time- and energy-efficient. This also saves working memory for other problems.

Yang and Schadlen [36] trained rhesus monkeys (*Macaca mulatta*) to combine probabilistic information from multiple sources to determine a correct outcome. The monkeys performed very well, weighing individual probabilities against each other. For example, a triangle occurs in 50% of the trials in which a red target is correct, but only in 5% in which a green target is correct, so they treated the triangle as a cue favouring the red target. A square occurs in 20% of red trials but in 30% of green trials, so squares favour green trials, but less strongly than triangles favour red targets. When presented with the simultaneous presentation of a triangle and a square, the monkeys correctly guessed red to be the most probable outcome. Interestingly, the authors measured the activation patterns of individual neurons in the monkeys' lateral intraparietal cortex and found that neural activity increased in accordance with the logarithm of the likelihood ratio. This research indicates that rhesus monkeys possess a specialized neural mechanism tuned to assessing probabilities. During a large part of our evolutionary history, humans depended on hunting and gathering. So we would also benefit from drawing statistical correlations during foraging. Yet, at first blush the human experimental data seem inconsistent with this prediction.

It begins to look like *Homo sapiens* does not deserve its epithet, but that bumblebees and other animals perform like good intuitive statisticians. One reason why bumblebees succeed where humans fail is that the former were tested in ecologically relevant conditions, whereas in aforementioned studies the latter are confronted with single case probabilities, which are less ecologically relevant [4]. Probability theorists disagree whether it is meaningful to assign probability to a single event. Bayesians deem this possible, because to them probability refers to a subjective degree of confidence. In contrast, frequentists argue that probability refers to the relative frequency with which an event occurs; therefore a single event cannot have a probability.

If we consider probability theory from an evolutionary perspective, it seems implausible that an organism could evolve that is capable of detecting single-event probabilities. It is impossible to state whether a single hunt in the northern canyon will be successful. On the other hand, encountered frequencies of actual events can be detected. Hunters can remember that 14 out of 20 hunts in the northern canyon were successful. On the basis of these and other evolutionary considerations, Cosmides and Tooby [6] predicted that people should be better at statistical reasoning when the problems are posed in frequentist terms. Indeed, Gigerenzer and Hoffrage [13] found that many of the apparent cognitive biases disappear when questions are rephrased in terms of natural frequencies. For example, medical experts made mistakes on the following breast cancer screening problem:

> The probability of breast cancer for women over 40 is 1%. The probability that someone who has breast cancer will test positive on a mammography is 80%; the probability that someone who has no breast cancer will test positive is 10%. Lucy, aged 42, tests positive. How much chance is there that she has breast cancer?

Gigerenzer and Edwards [10] presented physicians with this and similar statistical problems. Most physicians gave the wrong intuitive answer, 80%, which is far above the one yielded by Bayes' rule.

Applying Bayes' theorem, we calculate the hypothesis that Lucy has breast cancer as follows:

$$P(H|D) = \frac{P(H)P(D|H)}{P(H)P(D|H) + P(-H)P(D|-H)}$$

$$P(H|D) = \frac{(.01)(.80)}{(.01)(.80) + (.99)(.1)}$$

$$P(H|D) = .075$$

However, performance improved markedly once physicians were presented with the same problem in a frequency format:

> 10 out of 1000 women over 40 have breast cancer. 8 of those 10 with breast cancer will test positive; 99 of the 990 women without breast cancer will also test positive. How many of those with breast cancer will test positive?

This time, about half of the physicians arrived at the Bayesian response (about 8 %). Conversely, it is important to stress that without explicitly formulating problems as frequencies, naive people will make non-normative judgments. For example, a random survey of pedestrians in five metropolitan areas (New York, Amsterdam, Berlin, Milan and Athens) found that most people failed to understand what 'a 30% chance of rain tomorrow' means. Most subjects thought it meant that it would rain 30% of the time, followed closely by 'in 30% of the area', and even '30% of the metereologists believe it will rain tomorrow' [11]. The subjects answered incorrectly because the meteorologists have not specified the reference class to which probabilities of precipitation pertain (in this case, when weather conditions are like today, at least a minimal amount of rain will fall in 3 out of 10 cases the next day). Cosmides and Tooby [6] and Brase et al. [4] obtained similar findings with statistically naive subjects: problems posed in such a way that the relevant information is expressed as frequencies consistently elicit more normative responses than those that are formulated as percentages. Brase et al. [4] have therefore proposed an intimate relationship between probabilistic inference and our ability to count frequencies. Tentatively, it is interesting to note that the neurons involved in probability estimation in the rhesus monkey brain lie anatomically very close to number-sensitive neurons, which assess quantity [24]: both lie in the intraparietal area of the neocortex.

Taken together, these studies suggest that the apparent tension between intuitive probability and the rules and axioms of probability theory is largely resolved when subjects are presented with the problems in an ecologically relevant format. Even children as young as six years of age have been shown to solve complex probability problems, e.g., they preferred a less risky gamble with a smaller prize to a risky gamble with a higher prize [28]. Remarkably, the children could even multiply probabilities, although at this age they have not yet learned to multiply numbers. Given the survival advantages in terms of assessments of risks and benefits of organisms that can detect probabilities, it is perhaps not surprising that people are good intuitive statisticians. Indeed, some developmental psychologists working on language acquisition argue that infants rely on the statistical detection of recurring sound patterns to chunk streams of continuous speech into words, which is crucial for word learning [1]. In the following section, I examine whether frequency based intuitive probabilistic reasoning may be a reliable heuristic in scientific practice. I will discuss the role of the frequency heuristic in two case-studies from the history of biology, Darwin's origin of species, and Zahavi's handicap principle.

4   Scientific practice may be informed by intuitive probabilistic reasoning

Since scientists are equipped with the same cognitive capacities as other people, it is reasonable to expect that heuristics play an important role in scientific reasoning. Indeed, Wimsatt [35] argued that scientific virtues such as generality, simplicity, and fruitfulness can be thought of as heuristics with a normative force. Occam's razor is such a heuristic because it tells us to choose the simpler model or theory, all else being equal. In this way, it can help scientists decide between competing models which seem equally promising. Precisely because heuristics aim to reduce the complexity of a problem, they sometimes lead to errors. The history of science shows how the use of Occam's razor, for example, did not always lead to correct inferences: scientists who chose one theory over another because of its parsimony sometimes turned out to be wrong. In the history of biogeography, Croizat's theory that the dispersal patterns of animals (with some being very widespread and others restricted to particular continents or even islands) can be explained by tectonic change was rejected by many biologists, because they felt this theory to be less parsimonious than Wallace's straightforward explanation that the animals simply migrated. As it turned out, the less parsimonious theory was correct [2]. Of course, this does not imply that Occam's razor is unwarranted, but it does demonstrate that it is a heuristic that reduces the full complexity of a problem and therefore sometimes leads to incorrect conclusions.

Another example of heuristics in scientific practice is the debate between early twentieth-century evolutionary biologists on whether replicability or consilience was the most important virtue. Replicability was defended by biologists who placed a premium on experiments. Those who favoured consilience argued that as many lines of evidence as possible should be pursued. Both heuristics came with their own biases: the replicability heuristic considerably narrowed the range of evidence from which evolutionary biologists could draw. On the other hand, the consilience heuristic was limited by the knowledge of an individual naturalist, and could therefore affect the quality of the various lines of evidence being used [19]. Even though biology is now more driven by experiments, naturalists, like Ernst Mayr, who adhered to consilience as an epistemic virtue became the architects of the modern species concept.

Does intuitive probabilistic reasoning also play a role in scientific understanding? One way to approach this question is to examine the emergence of probability and statistical inference in the history of science and its current use in scientific practice. Hacking [16] distinguished between two distinct concepts of probability: on the one hand there is the degree of belief warranted by the evidence (epistemological probability), on the other hand there is the tendency, displayed by some chance devices, such as dice, to produce stable relative frequencies (statistical probability). In the history of western thought, the epistemological concept predates the statistical view by several centuries: during the Middle Ages, probability was an evaluation of opinion — opinions in this view were probable when they were approved by authority, or supported by ancient or holy books [16, p. 30]. With the emergence of early modern science in the sixteenth century, the observation of empirical phenomena became more important than testimony. However, these observations varied in reliability, and there was no external justification in terms of testimony by authorities to weigh conflicting evidence. Statistical probability, developed by Leibniz and others, defined probability as the ratio between equally possible events. Interestingly, as Ritchie [27] argues, this latter form of probability is sometimes used by both scientists and laypeople to express their subjective degree of confidence warranted by the evidence. In this case, statistical probabilities are used to express a (subjective) degree of belief warranted by the evidence. For a scientific example, take a randomized medical controlled trial that shows that 30 people from a group of 100 who received a vaccine got the disease, compared to 60 subjects in the control group of 100 subjects who only received a placebo. To examine the effectiveness of the vaccine, we test the null hypothesis, i.e., how likely is it that 60 of the 90 people who got the disease in any event would be assigned to the control group? It turns out to be very low: $p < .001$ (χ^2 test, 95% confidence interval). In other words, if the null hypothesis is

true, then we expect to observe such a large difference at the most once in
a thousand times. This result is statistically significant, i.e., the researcher
can be confident to conclude from this clinical trial that the vaccine is most
probably effective. In this way, statistical probability, as expressed in terms
of statistical significance, has become a metaphor for epistemological probability in scientific understanding. Interestingly, the submission guidelines
of major medical journals (such as British Medical Journal) include the
warning not to confuse statistical significance with *scientific* significance,
which indicates that researchers indeed often use statistical probability as
a measure of epistemological probability. We also use statistical probability
in this sense in everyday language, as in Jeff Wayne's musical renditioning
(1978) of H. G. Wells's *War of the Worlds*, 'The chances of anything coming from Mars are a million to one'. This use of statistical probability as a
measure of epistemological probability can be explained by the fact that our
cognitive architecture was shaped by natural selection to deal with natural
frequencies. We find it more intelligible and easier to think of probabilities
as frequencies. In the following case-studies, I show that intuitive statistical
probability can serve as a source of scientific creativity and understanding.

4.1 On the Origin of Species

In his *On the origin of species* [9], Charles Darwin set out an informal,
qualitative probabilistic argument to defend his idea that species evolve
through the principle of natural selection. Statistical methods to examine
the effects of natural selection were not available at that time — indeed,
upon acceptance of the principle of natural selection it became an important
challenge to describe it in mathematical terms. Given that Darwin had no
access to mathematical models to test the plausibility of his theory, how
could he have developed it? Several philosophers of biology argue that
population thinking may have been the key. Ernst Mayr [21, p. 47] goes as
far as to say that 'Darwin would not have arrived at a theory of natural
selection if he had not adopted population thinking'. In population thinking,
biologists examine species not as idealized classes of entities, but as groups
of individuals which differ in their ability to survive and reproduce. This
way of thinking enabled Darwin to use intuitive frequentist probabilistic
inferences.

The theory of natural selection can be summarized as follows. If resources are limited, and if a population can produce more offspring than
there are resources, competition will ensue. If individuals within the population exhibit variation, then some will be more likely to survive and reproduce than others. If these variations are heritable, they will become fixed in
the population, which as a result will become more adapted to the environment. It is difficult to get an intuitive grasp of Darwin's theory of natural

selection by assigning probabilities to single events. A well-known thought experiment involves a creature, perfectly adapted in every sense, which unfortunately gets struck by lightning before it reaches reproductive age. It is more sensible to think about natural selection in frequentist terms. Darwin therefore referred to populations, rather than individuals. This is why he adopted the term 'survival of the fittest' (invented by Herbert Spencer) in the later editions of the *Origin of Species*, alongside his own term 'natural selection', precisely because the former puts more emphasis on the relatively higher survival rate of the most adapted individuals within a population. Consider this example from chapter 4:

> *Illustrations of the action of Natural Selection* — In order to make it clear how, as I believe, natural selection acts, I must beg permission to give one or two imaginary illustrations. Let us take the case of a wolf, which preys on various animals, securing some by craft, some by strength, and some by fleetness; and let us suppose that the fleetest prey, a deer for instance, had from any change in the country increased in numbers, or that other prey had decreased in numbers, during that season of the year when the wolf is hardest pressed for food. I can under such circumstances see no reason to doubt that the swiftest and slimmest wolves would have the best chance of surviving, and so be preserved or selected [9, p. 90-91].

It is not very meaningful to think of an individual organism under difficult circumstances, given a particular advantageous trait, to have a 60 % chance of survival, compared to a 40 % base rate! From Darwin's books, notebooks and letters (cf., e.g., [23] for a detailed examination), we know that the 1798 work of Thomas Malthus [20] was critical for this change to population thinking. Malthus' examination of the struggle of existence in humans enabled Darwin to see competition not only between species, but also between members of the same species. In the introduction of the *Origin of Species*, Darwin makes this link explicit:

> This is the doctrine of Malthus, applied to the whole animal and vegetable kingdoms. As many more individuals of each species are born than can possibly survive; and as, consequently, there is a frequently recurring struggle for existence, it follows that any being, if it vary however slightly in any manner profitable to itself, under the complex and sometimes varying conditions of life, will have a better chance of surviving, and thus be naturally selected. From the strong principle of inheritance, any selected variety will tend to propagate its new and modified form [9, p. 5].

Darwin's use of frequencies is also apparent in his personal notebooks, where he experimented with the idea that external environmental constraints can provide enough selection pressures to change species. Although he knew that artificial selection (e.g., selective breeding) could result in some minor changes (such as tail shape in domestic pigeons), he needed a strong force to explain how natural selection could result in speciation.

> One may say there is a force like a hundred thousand wedges trying [to] force every kind of adapted structure into the gaps in the oeconomy of nature or rather forming gaps by thrusting out weaker ones. [...] The final cause of all this wedging, must be to sort out proper structure, & adapt it to changes [8].

In this case, it is not the individuals but the selective pressures themselves that are being expressed in frequentist terms.

Prior to Darwin, biologists did not consider populations but focused on individuals. Linnaeus and others made standard idealized descriptions of species of animals and plants such as 'the daisy', 'the honeybee'. It seems virtually impossible that taxonomists should have overlooked the individual variation within species, given that they studied dozens of specimens before making a detailed description of a given species. Well into the 19th century, virtually all biologists were species essentialists: they believed animals and plants to possess an underlying species-typical essence, which guided their behaviour and development [31]. The foundations for this essentialism are twofold. First, it has ancient philosophical roots in Plato and the Pythagoreans, who postulated that the world consists of a limited number of classes (*eide*) and that only the essence of each class has reality. Second, essentialism seems to be a stable and universal part of human cognition. Even children as young as three or four years, from backgrounds as diverse as Western Africa, Mongolia or Western Europe, believe that living things possess an unchanging hidden essence [7]. This essentialist stance led pre-Darwinian taxonomists to ignore or downplay the natural variability that species exhibit.

When one considers individual organisms, it is impossible to arrive at the theory of natural selection. William Paley, for example, in his *Natural Theology* [25] discussed individual adaptations within specific organisms. The beginning of *Natural Theology* is diagnostic of this individual based approach. In his well-known watch analogy, Paley invites us to consider the improbability that an individual watch could come about by chance. Without taking into account the variation on the population level, it is indeed difficult to imagine how exquisite adaptations such as the eye could have evolved. The theory of natural selection resisted mathematical description for over half a century. It was only in the early 1920s that population biologists such as Ronald A. Fisher developed statistical methods that could discriminate the effects of natural selection from other evolutionary processes.

4.2   Zahavi's handicap principle

Despite this formalization, intuitive probabilistic reasoning continues to play a role in evolutionary theory, as I shall illustrate with my second case-study, the handicap principle, first formulated by Amotz Zahavi in 1975 [37]. In brief, it states that high-quality animals may benefit from advertising their qualities to potential mates, competitors or predators. They take on a handicap, in a way that inferior individuals would not be able to do, because for the latter the investment would be too high. A healthy gazelle stots (makes high leaps) in front of a charging lioness instead of running away, to convince her that a pursuit would be a loss of time. The lioness desists, and will instead attack another gazelle that does not stot and that is likely to be weaker. The peacock's tail clearly hinders its owner to forage in the undergrowth, or to escape from predators, and in weaker individuals it attracts parasites. Peahens, therefore, use the peacock's tail as an honest signal of health, and consistently choose males with longer tails [39]. Although this model elegantly explains the existence of seemingly redundant and costly traits and behaviours, it was not accepted by the majority of biologists, who distrusted it because it was not stated in a mathematical form. As Zahavi [38, p. 860] later put it 'For some reason that I cannot understand, logical models expressed verbally are often rejected as being "intuitive"'.

In 1990, Alan Grafen in two landmark papers [14, 15] finally made the handicap principle acceptable for biologists by reformulating it in a game theoretical model. However, Grafen stated that his conclusions were the same as in Zahavi's original paper [15, p. 487], and he admitted that the model was simple enough to be stated verbally [14, p. 541]. Nevertheless, according to Zahavi 'biologists remained unimpressed by the logic of the verbal model, and accepted the handicap principle only when expressed in a complex mathematical model, which I and probably many other ethologists do not understand' [38, p. 860].

How then was Zahavi able to craft the handicap principle without mathematical modeling? It is difficult to intuitively examine the handicap principle in single event probabilities: imagine a gazelle that stots before a hungry lioness in an attempt to honestly show that pursuing it would be a waste of time and energy. This single event has as outcome that the lioness either calls its bluff by pursuing it or not. What happens when we examine the handicap principle in frequentist terms? Given many generations of gazelles that honestly signaled their speed, lionesses that did pursue them had a lower rate of success —since the signal is honest— while those that focused on non-stotting gazelles had more success. From the perspective of the gazelles, those that stotted wasted less energy and time, as stotting is less costly than running for one's life — whereas those that did not signal their speed wasted more energy. From the perspective of the lionesses, those that accepted stotting as an honest signal, wasted less energy and time pursuing prey that was too fit to kill. Early in his observational studies of birds, Zahavi looked at benefits for both signalers, who display the handicap, and the ones they signal to, the predators or potential mates who obtain information about the signalers through their handicap. From a frequentist point of view, the evolution of handicaps becomes less puzzling; it is easier to see how such behaviour could have evolved.

5 Conclusion

The received view in cognitive science holds that intuitive probabilistic reasoning is biased or even fatally flawed. However, this conclusion is based on experimental paradigms where subjects are asked to assign probabilities to single events. As single events are not meaningful from an evolutionary point of view, the evolved structure of the human mind is poorly equipped to handle them. Experiments in which people are asked to assign probabilities to natural frequencies are more ecologically relevant, and hence yield better performances. Humans share with other animals an intuitive sense of frequentist reasoning, which is more ancient than both philosophical and scientific concepts of probability. Converging evidence from experimental, developmental and comparative psychology suggest that this intuitive probability is more or less in line with the rules and axioms of formal probability theory. In this paper, I have argued that intuitive frequentist reasoning plays a productive role in scientific understanding. I have shown that both Darwin and Zahavi relied on naive frequentist reasoning to develop their models for which the mathematics was lacking. This lends support to the hypothesis that efficient heuristics can guide scientists in the early creative stages of theory formation. Given that scientists draw on the same cognitive resources as other people, this finding may not be as surprising or counter-intuitive as it may seem at first.

References

[1] Richard N. Aslin, Jenny R. Saffran, and Elissa L. Newport. Computation of Conditional Probability Statistics by 8-month-old Infants. *Psychological Science*, 9:321–324, 1998.

[2] Alan Baker. Occam's Razor in Science: a Case Study from Biogeography. *Biology and Philosophy*, 22:193–215, 2007.

[3] Melissa Bateson, Susan D. Healy, and T. Andrew Hurley. Context-dependent Foraging Decisions in Rufous Hummingbirds. *Proceedings of the Royal Society London B*, 270:1271–1276, 2003.

[4] Gary L. Brase, Leda Cosmides, and John Tooby. Individuation, Counting, and Statistical Inference: The Role of Frequency and Whole-object Representation in Judgment under Uncertainty. *Journal of Experimental Psychology General*, 127:3–21, 1998.

[5] Colin F. Camerer. Bounded Rationality in Individual Decision Making. *Experimental Economics*, 1:163–183, 1998.

[6] Leda Cosmides and John Tooby. Are Humans Good Intuitive Statisticians after all? Rethinking Some Conclusions from the Literature on Judgment under Uncertainty. *Cognition*, 58:1–73, 1996.

[7] Helen De Cruz and Johan De Smedt. The Role of Intuitive Ontologies in Scientific Understanding—the Case of Human Evolution. *Biology and Philosophy*, 22:351–368, 2007.

[8] Charles Darwin. Notebook D, 1844. unpublished manuscript, online at darwin-online.org.uk.

[9] Charles Darwin. *On the Origin of Species by Means of Natural Selection or the Preservation of Favoured Races in the Struggle for Life*. London, 1859.

[10] Gerd Gigerenzer and Adrian Edwards. Simple Tools for Understanding Risks: from Innumeracy to Insight. *British Medical Journal*, 327:741–744, 2003.

[11] Gerd Gigerenzer, Ralph Hertwig, Eva van den Broek, Barbara Fasolo, and Konstantinos V. Katsikopoulos. A 30% Chance of Rain Tomorrow: how Does the Public Understand Probabilistic Weather Forecasts? *Risk Analysis*, 25:623–629, 2005.

[12] Gerd Gigerenzer and Ulrich Hoffrage. Bayesian Reasoning without Instruction: Frequency Formats. *Psychology Review*, 102:684–704, 1995.

[13] Gerd Gigerenzer and Ulrich Hoffrage. Overcoming Difficulties in Bayesian Reasoning. *Psychological Review*, 106:425–430, 1999.

[14] Alan Grafen. Biological Signals as Handicaps. *Journal of Theoretical Biology*, 144:517–546, 1990.

[15] Alan Grafen. Sexual Selection Unhandicapped by the Fisher Process. *Journal of Theoretical Biology*, 144:473–516, 1990.

[16] Ian Hacking. *The Emergence of Probability: A Philosophical Study of Early Ideas about Probability, Induction and Statistical Inference.* Cambridge University Press, 2nd edition, 2006.

[17] Philip N. Johnson-Laird, Paolo Legrenzi, Vittorio Girotto, and Maria S. Legrenzi. A Mental Model Theory of Extensional Reasoning. *Psychological Review*, 106:62–88, 1999.

[18] Daniel Kahneman and Amos Tversky. On the Reality of Cognitive Illusions. *Psychological Review*, 103:582–591, 1996.

[19] David Magnus. Heuristics and Biases in Evolutionary Biology. *Biology and Philosophy*, 12:21–38, 1997.

[20] Thomas Malthus. *An Essay on the Principle of Population.* London, 1798.

[21] Ernst Mayr. *The Growth of Biological Thought.* Belknap Press, 1982.

[22] Michael McCloskey, Alfonso Caramazza, and Bert Green. Curvilinear Motion in the Absence of External Forces: Naive Beliefs about the Motion of Objects. *Science*, 210:1139–1141, 1980.

[23] Arthur B. Millman and Carol L. Smith. Darwin's Use of Analogical Reasoning in Theory Construction. *Metaphor and Symbol*, 12:159–187, 1997.

[24] Andreas Nieder and Earl K. Miller. A Parieto-Frontal Network for Visual Numerical Information in the Monkey. *Proceedings of the National Academy of Sciences of the United States of America*, 101:7457–7462, 1991.

[25] William Paley. *Natural Theology or Evidence of the Existence and Attributions of the Deity Collected from the Appearances of Nature.* Oxford World's Classics, 2006. Original edition 1802.

[26] Leslie A. Real. Animal Choice Behavior and the Evolution of Cognitive Architecture. *Science*, 253:980–986, 1991.

[27] David L. Ritchie. Statistical Probability as a Metaphor for Epistemological Probability. *Metaphor and Symbol*, 18:1–11, 2003.

[28] Anne Schlottmann. Children's Probability Intuitions: Understanding the Expected Value of Complex Gambles. *Child Development*, 72:103–122, 2001.

[29] Shinsuke Shimojo and S. Ichiwaka. Intuitive Reasoning about Probability: Theoretical and Experimental Analyses of the "Problem of Three Prisoners". *Cognition*, 32:1–24, 1989.

[30] Herbert Simon. Rational Choice and the Structure of the Environment. *Psychological Review*, 63:129–138, 1956.

[31] David N. Stamos. Pre-Darwinian Taxonomy and Essentialism. *Biology and Philosophy*, 20:79–96, 2005.

[32] Keith E. Stanovich and Richard F. West. Individual Differences in Reasoning: Implications for the Rationality Debate? *Behavioral and Brain Sciences*, 23:645–726, 2000.

[33] Amos Tversky and Daniel Kahneman. Judgment under Uncertainty: Heuristics and Biases. *Science*, 185:1124–1131, 1974.

[34] Cheryl J. Wakslak, Yaacov Trope, Nira Liberman, and Rotem Alony. Seeing the Forest when Entry is Unlikely: Probability and the Mental Representation of Events. *Journal of Experimental Psychology General*, 135:641–653, 2006.

[35] William Wimsatt. Reductionistic Research Strategies and Their Biases in the Units of Selection Controversy. In Thomas Nickles, editor, *Scientific Discovery: Case Studies*, volume 60 of *Boston Studies in the Philosophy of Science*, pages 213–259. Reidel, 1980.

[36] Tianming Yang and Michael N. Shadlen. Probabilistic Reasoning by Neurons. *Nature*, 447:1075–1080, 2007.

[37] Amotz Zahavi. Mate Selection: A Selection for a Handicap. *Journal of Theoretical Biology*, 53:205–214, 1975.

[38] Amotz Zahavi. Indirect Selection and Individual Selection in Sociobiology: My Personal Views on Theories of Social Behaviour. *Animal Behaviour*, 65:859–863, 2003.

[39] Amotz Zahavi and Avishag Zahavi. *The Handicap Principle: A Missing Piece of Darwin's Puzzle*. Oxford University Press, 1977.

Benedikt **Löwe**, Eric **Pacuit**, Jan-Willem **Romeijn** (*eds.*)
Foundations of the Formal Sciences VI
Reasoning about Probabilities and Probabilistic Reasoning

Can there be a propensity interpretation of conditional probabilities?

ISABELLE DROUET*

Institut d'Histoire et de Philosophie des Sciences et des Techniques (IHPST), Université Paris 1 Panthéon-Sorbonne, 13, rue du Four, 75006 Paris, France
E-mail: isabelle.drouet@gmail.com

1 Introduction

The propensity interpretation of probability was introduced by Popper during the late 1950s (cf. [9, 10]). Its main appeal was (and still is) to let singular probabilities be physical — that is dependent only on the state of the world. Indeed, the fundamental idea behind the interpretation is that probabilities of singular events depend on the physical system that produces those events. For example, the probability of getting a "six" on next throw of a given die depends on the physical properties of the throwing device — i.e., in the present case, mainly the physical properties of the die itself and of the surface on which it is to be thrown. Due to its physical properties, the throwing device has a propensity to produce a "six" on next throw, and the probability of getting a "six" on next throw measures this propensity. In short, probabilities measure propensities to produce singular events. To the extent that those propensities differ from physical system to physical system, probabilities depend on the physical system one considers.

Since its introduction, the propensity interpretation of probability has been confronted with many criticisms. Among those criticisms, the most robust as well as central one is undoubtedly "Humphreys' paradox". According to this criticism, there cannot be a propensity interpretation of conditional probabilities. To put it in a slightly different way, conditional

---

*I thank participants of the "Probability, Decision, Uncertainty" seminar at IHPST (Paris) where I first presented this work. I am also very grateful to Paul Humphreys for his patient and kind help, and to two anonymous referees for careful reading and insightful comments.

Received by the editors: 29 September 2007; 15 June 2008.
Accepted for publication: 16 October 2008.

propensities would fail to be conditional probabilities. Two important consequences ensue. In the first place, one has to give up the idea of a correspondence between subjective and physical probabilities that would conform
to Lewis' *Principal Principle*. Roughly, the *Principal Principle* states that
a rational agent should believe to the degree x in A when knowing that
physical probability of A is x.[1] Since degrees of rational conditional belief
are conditional probabilities [13], the *Principal Principle* is not compatible
with conditional propensities failing to be conditional probabilities. In the
second place, one has to give up the idea that probability theory as we know
it is the theory of all aleatory events.

Now, the present paper calls Humphreys' conclusion into question, and
actually suggests a propensity interpretation of conditional probabilities.
The originality of the proposed solution lies less in its content, than in the
way it is approached. More precisely, the solution is built out of an analysis
of what it means to interpret conditional probabilities. Accordingly, the first
section of the paper is a presentation of Humphreys' paradox, the second
section deals with the very notion of interpreting conditional probabilities
and the third section sets out a solution to Humphreys' paradox. Finally, a
fourth section discusses the proposed solution.

2   Humphreys' paradox

As noticed by Humphreys himself, there exists a "variety of versions [5, p.
668]" of the paradox. Some of these versions are formal and some are not.
Informal versions are often vague. In general, it is clear neither what they
exactly consist in, nor whether there exists an equivalent formal version of
the difficulty. Therefore, I focus on formal versions: as Humphreys points
out, "for those, a satisfactory solution is required [5, p. 668]". To be more
precise, I first focus on the formal paradox as it stands in Humphreys'
seminal paper [4].[2] Once this is done, I explain how the initial paradox can
be generalized, giving rise to other formal versions of the paradox.

Both Humphreys' original argument and its generalization (in [5]) depend on considering the physical system that is described as follows:

> Take, then, the case of a well-known physical phenomenon, the trans-
> mission and reflection of photons from a half-silvered mirror. A
> source of spontaneously emitted photons allows the particles to im-
> pinge upon the mirror, but the system is so arranged that not all the

---

[1]Cf. [7, p. 266]. The principle is in fact noticeably subtler than that. In particular, it
stipulates that other evidence that the agent might have be "admissible" at the time the
physical probability of A is x. However, these subtleties need not concern us here.

[2]The paper is seminal to Humphreys' paradox in the sense that it is the first paper
Humphreys devotes to his objection. Yet he already had the objection and the objec-
tion was already known of philosophers of science by the end of the 1970s. The name
"Humphreys' paradox" was introduced in [2].

photons emitted from the source hit the mirror, and it is sufficiently
isolated that only the factors explicitly mentioned here are relevant.
Let I_{t_2} be the event of a photon impinging upon the mirror at time
t_2, and let T_{t_3} be the event of a photon being transmitted through
the mirror at time t_3 later than t_2. Now consider the single-case
conditional propensity $\Pr_{t_1}(\_|\_)$ where t_1 is earlier than t_2. [4, p.
561]

This description implies the following three (in)equalities involving probabilities relative to the system under consideration[3]:

$$\Pr_{t_1}(T_{t_3}|I_{t_2}) = p > 0, \tag{1a}$$

$$1 > \Pr_{t_1}(I_{t_2}) = q > 0, \tag{1b}$$

$$\Pr_{t_1}(T_{t_3}|\text{not-}I_{t_2}) = 0. \tag{1c}$$

Intuitively, 1a expresses the fact that a photon impinging upon the mirror at
t_2 *can* be transmitted through the mirror at t_3, 1b the fact that an emitted
photon *may* impinge upon the mirror at t_2, 1c the fact that a non-impinging
photon *cannot* be transmitted. 1a to 1c are nothing but some aspects of
the characterization of Humphreys' example put in a slightly formal way.

Besides the (in)equalities just set out, Humphreys accepts the following:

$$\Pr_{t_1}(I_{t_2}|T_{t_3}) = \Pr_{t_1}(I_{t_2}|\text{not-}T_{t_3}) = \Pr_{t_1}(I_{t_2}). \tag{2}$$

Contrary to 1a to 1c, double equality 2 does not stem from the description
of the photon-emitting system. Rather, it is a substantial hypothesis concerning the way a specific probability should be evaluated. More precisely,
$\Pr_{t_1}(I_{t_2}|T_{t_3})$ has the specificity to be an inverse conditional probability, that
is a conditional probability such that the conditioning event (T_{t_3} in the case
in point) is posterior to conditioned event (I_{t_2}). Correlatively, it is not immediately clear at all how it should be evaluated when given a propensity
interpretation.[4] Indeed, propensities are causal-like entities, and therefore
what they become in time-reversed contexts is problematic.

Humphreys' answer concerning the evaluation of $\Pr_{t_1}(I_{t_2}|T_{t_3})$ can be
seen as an instantiation of a more general principle for the evaluation of
inverse conditional probabilities:

> If $\Pr(A|B)$ is an inverse conditional probability
> given a propensity interpretation, then $\Pr(A|B) =$ (CI)
> $\Pr(A|\text{not-}B) = \Pr(A)$.

---

[3]Cf. [4, p. 561]. For reasons that should become clear later on, my notation is slightly
different from Humphreys'. Let me also indicate that not-E refers to the event of event
E not occurring.

[4]Humphreys does not distinguish between propensities and probabilities interpreted
as propensities. For clarity's sake, I shall distinguish between them.

Humphreys' justification for principle (CI) is that posterior events do not (at least normally and in the situations that he addresses and that I also will address in the present text) influence prior events, and hence cannot modify the propensity for the system to produce those prior events. In particular, "the propensity for a particle to impinge upon the mirror is unaffected by whether the particle is transmitted or not. [4, p. 561]"

The argument provided by Humphreys in favor of accepting the principle (CI) is convincing. The matter is that (CI) is incompatible with some very basic properties of standard conditional probabilities. More precisely, Humphreys shows that hypotheses 1a to 1c and 2 taken together are compatible neither with Bayes' theorem, nor with what I shall call (following Humphreys) the "multiplication principle":

$$\Pr(AB) = \Pr(A|B) \cdot \Pr(B) = \Pr(B|A) \cdot \Pr(A) = \Pr(BA). \qquad \text{(MP)}$$

(MP) is an immediate consequence of the very usual "definition" of conditional probabilities as ratios of absolute probabilities. As a consequence, I shall focus here on the incompatibility of this principle with hypotheses 1a to 1c and 2.

In order to explain where the incompatibility comes from, one must first notice that the multiplication principle together with the additivity axiom for probabilities entail the law of total probability:

$$\Pr(A) = \Pr(A|B) \cdot \Pr(B) + \Pr(A|\text{not-}B) \cdot \Pr(\text{not-}B). \qquad \text{(TP)}$$

This having been recalled [4, p. 560], Humphreys computes $\Pr_{t_1}(I_{t_2}T_{t_3})$ in two different ways and notices that the results of these different computations are incompatible. On the one hand, from (MP) and hypotheses 1a and 1b, we have: $\Pr_{t_1}(I_{t_2}T_{t_3}) = \Pr_{t_1}(T_{t_3}I_{t_2}) = \Pr_{t_1}(T_{t_3}|I_{t_2})\Pr_{t_1}(I_{t_2}) = pq$. On the other hand, (MP) implies that $\Pr_{t_1}(I_{t_2}T_{t_3}) = \Pr_{t_1}(I_{t_2}|T_{t_3})\Pr_{t_1}(T_{t_3})$. Following 2 and 1b, we have $\Pr_{t_1}(I_{t_2}|T_{t_3}) = \Pr_{t_1}(I_{t_2}) = q$. From (TP) followed by substitutions with the values from 1a to 1b, we get

$$
\begin{aligned}
\Pr_{t_1}(T_{t_3}) \;&=\; \Pr_{t_1}(T_{t_3}|I_{t_2})\,\Pr_{t_1}(I_{t_2}) \\
&\quad + \Pr_{t_1}(T_{t_3}|\text{not-}I_{t_2})\,\Pr_{t_1}(\text{not-}I_{t_2}) \\
&=\; pq + 0 = pq.
\end{aligned}
$$

As a consequence, $\Pr_{t_1}(I_{t_2}T_{t_3}) = pq^2$. We then have $pq = pq^2$, which is incompatible with the conjunction of 1a and 1b. This is Humphreys' paradox in regard to (MP). Humphreys concludes that conditional probabilities cannot receive a propensity interpretation.

As already stated, 1a to 1c belong to the very characterization of the photon-emitting system considered by Humphreys. On the other hand,

(MP) and the additivity axiom (which, together with (MP), enables to derive (TP)) give fundamental, unescapable properties of standard conditional probabilities. Consequently, rejecting (CI) appears as the only strategy out of Humphreys' paradox. Positively, one may substitute for it another principle for evaluating inverse conditional probabilities under a propensity interpretation. Apart from (CI), the literature on the propensity interpretation has two such principles to offer, viz.

$$\text{If } \Pr(A|B) \text{ is an inverse conditional probability given a propensity interpretation, } \Pr(A|B) = 0 \qquad \text{(ZI)}$$

and

$$\text{if } \Pr(A|B) \text{ is an inverse conditional probability given a propensity interpretation, } \Pr(A|B) = 0 \text{ or } \qquad \text{(FP)}$$
$$\Pr(A|B) = 1.$$

The principle (ZI), defended in particular by Fetzer [2], expresses the fact that the influence of a posterior event on a prior event is null. (FP), defended in particular in [8], expresses the fact that, at the time corresponding to B, A has definitely occurred or failed to occur.

Obviously, it should be debated which one of principles (CI), (ZI) and (FP) is adequate for evaluating inverse conditional probabilities when they are given a propensity interpretation. However, whatever the conclusion of such a debate may be, it will not be such that the propensity interpretation suits conditional probabilities. More precisely, it is claimed in [5] that there exist formal versions of the paradox analogous to the initial one, but dealing with (ZI) for the first one and (FP) for the second one [5, p. 670]. These versions of the paradox arise along the same lines as the initial one. The difference from the original argument occurs when $\Pr_{t_1}(I_{t_2}|T_{t_3})$ is computed out of 2. The problem, now, is exactly that (ZI) and (FP) may impose that this probability is 0.[5] Indeed, if $\Pr_{t_1}(I_{t_2}|T_{t_3}) = 0$, then the second way of computing $\Pr_{t_1}(I_{t_2}T_{t_3})$ leads to 0. This should be equal to pq that is given by the first computation, but this cannot be the case since both p and q differ from 0 by hypothesis. Through these (ZI) and (FP) versions, Humphreys' paradox is generalized.

At that point, the situation may look quite desperate. Principles for the evaluation of inverse conditional probabilities in a propensity context have all revealed incompatible with the fundamental property of conditional probabilities that is expressed by (MP). Still, two remarks can be made by one who would like to give a propensity interpretation of conditional probabilities. To begin with, it must be emphasized that none of the proponents of principles (CI), (ZI) or (FP) for the evaluation of inverse conditional

---

[5]This is always the case according to (ZI), and this is the case whenever I_{t_2} fails to occur according to (FP).

propensities explicitly raises the question of how conditional probabilities should be interpreted in the propensity framework. To put it another way, they implicitly agree on the idea that the propensity interpretation of absolute probabilities analytically contains an interpretation of conditional probabilities, meaning that what conditional probabilities should be taken as measuring deductively stems from what absolute probabilities measure. But the very debate reveals that this assumption is false. Indeed, diverging principles for the evaluation of inverse conditional probabilities rely on different conceptions of the way conditional probabilities should be interpreted in the propensity framework. More precisely, Humphreys' (CI) cannot be separated from the idea that $\Pr(A|B)$ measures the propensity that tends to realize A in as much as it is possibly physically modified by the occurrence of B, (ZI) presupposes that $\Pr(A|B)$ measures the causal influence of B on A, and (FP) stems from the idea that this very probability measures the propensity tending to realize A as it stands at the moment which is characteristic of B. Such a disagreement would not happen if it were true that the propensity theory of absolute probabilities analytically contains an interpretation of conditional probabilities. As a consequence, it should and will be considered an open question how conditional probabilities could be interpreted in a propensity framework.

In this context, what [4] and [5] show is only that a certain number of answers will not do because they lead to principles for the evaluation of inverse conditional probabilities that are not compatible with standard probability theory. But, my second remark goes, one must notice that Humphreys has not given any general argument against the very idea of producing a propensity interpretation of conditional probabilities. If the question of how conditional probabilities should be interpreted in a propensity framework is indeed an open question, [4] and [5] do not rule out the possibility of this question receiving a satisfactory answer, an answer that would be compatible with the usual properties of conditional probabilities. In order to reach such an answer, I suggest that we start with discussing the very notion of interpreting conditional probabilities.

3   Interpreting conditional probabilities

The focus in this section is on what it is to *interpret* conditional probabilities, rather than on what conditional probabilities *are*. Granted, the two questions are hardly independent. More precisely, it seems to be a natural requirement that one knows what conditional probabilities are before one tackles the question of how they should be interpreted. However, as natural as it may seem, this line of thought is too stringent to have been followed in practice. On the one hand, there already exist philosophical interpretations of conditional probabilities (think subjective, frequentist or logical). On the other hand, it remains an unsettled matter what conditional probabilities really are or how they should be defined.[6] I shall not try to settle this matter before I turn to the question of how conditional probabilities should be interpreted in a propensity framework. I shall only take it that conditional probabilities satisfy the usual characterization as ratios of absolute probabilities whenever the denominator is not zero, but I shall not pronounce myself on how conditional probabilities should be defined exactly.[7] This is justified by current state of the art on interpreting conditional probabilities. In the propensity context, we do not even have the slightest idea of what such an interpretation could be. Hence I would say that accounting for cases when the denominator of the ratio characterization is zero should not considered an absolute priority.

Coming to the question of what it is to interpret conditional probabilities, I start with the following observation: the subjectivist account provably succeeds in providing an interpretation of both absolute and conditional probabilities.[8] My strategy, thus, is as follows: examine the way subjectivists interpret conditional probabilities and draw from this a general proposition as to what it is to interpret conditional probabilities. Now, under the subjectivist interpretation, absolute probabilities measure degrees of rational belief. Specifically, let Pr be the function measuring the degrees of rational belief of individual I under the stock of information K. Then, the subjectivist claims, $\Pr(\_|A)$ is the function measuring I's degrees of rational belief under information $K \cup \{A\}$. In other words, conditionalizing on A amounts to adding A to I's initial stock of information. This makes two points: first that conditionalization indeed has to be interpreted, second that the interpretation consists in explicating how an initial probability function (here: the probability function corresponding to I's degrees of belief under stock of information K) is modified by the conditioning element (proposition A).

The subjectivist illustration allows to go a little further in the analysis what of it is to conditionalize — and, consequently, to interpret conditionalization. Indeed, it is clear from the presentation just given that a probability function interpreted in the subjectivist way depends on the individual I one considers, as well as on I's stock of information K. Since I and K determine the probability function, they could be called its "determinants". They differ from the arguments of the function which, in the subjectivist case, are propositions. Now it appears that interpreting conditionalization consists of telling how an argument redefines the determinant of the prob-

---

[6]Two recent papers attesting to the fact that the matter remains unsettled are [3, 1].

[7]Still, I would claim that the propensity interpretation would better suit a definition that makes absolute probabilities primitive, very much along the lines followed in [1].

[8]This is established in [11] for the absolute case and in [13] for the conditional one.

ability function. More rigorously, conditionalization is to be interpreted as a function which associates a new determinant to any pair composed of an initial determinant and an argument. Let us put it formally in the subjectivist case. To that effect, let $\mathbf{I}$ be the sets of individuals, $\mathbf{P}$ the set of propositions, and $\mathbf{K}$ the powerset of $\mathbf{P}$. Then the subjectivist interpretation of conditionalization is by the function c_s defined as follows:

$$
\begin{aligned}
c_s \quad : \quad & (\mathbf{I} \times \mathbf{K}) \times \mathbf{P} \quad \longrightarrow \quad \mathbf{I} \times \mathbf{K} \\
& ((I, K), P) \quad \longmapsto \quad (I, K') = (I, K \cup \{P\}).
\end{aligned}
$$

What is established in [13] is that c_s so defined accounts for the properties of Bayesian conditionalization.

This leads to a general analysis of an interpretation of probability as consisting of:

1. an interpretation of absolute probabilities. This must specify in particular:

    (a) what kind of objects arguments of probability functions are;

    (b) what kind of objects determinants of probability functions are;

2. an interpretation of conditionalization as a function from the cartesian product of the set of arguments and the set of determinants, to the set of determinants.

This analysis is only about the kind of thing an interpretation of probability is. It does not prejudge how absolute probabilities and conditionalization should actually be interpreted. Even less does it mean that any proposition for the nature of arguments and determinants, or any function from the cartesian product of the set of arguments and the set of determinants to the set of determinants will do. Quite the opposite, it is clear that most of the possible propositions will fail to be admissible.[9] But armed with this analysis, I can come back to the propensity interpretation and try to produce an interpretation of conditional probabilities that will actually be admissible.

4   A propensity interpretation of conditional probabilities

It is clear from the presentation given in the introduction that under the propensity interpretation of probability:

1. absolute probabilities measure the propensities of physical systems to realize singular events. Hence,

---

[9] "Admissible" can be taken here in the sense introduced in [12, pp. 63–64].

(a) arguments of probability functions are elements of the set **E** of singular events. Following the usual mathematical representation, **E** is a sigma-algebra;

(b) determinants of probability functions are elements of the set **S** of physical systems.

Consequently,

2. conditionalization is to be interpreted as a function c_p from the cartesian product of the set of physical systems with the set of singular events, to the set of physical systems.

At that point, the question of the propensity interpretation of conditional probabilities becomes the question of defining c_p. In other words, analyzing what it is to interpret conditionalization leads us to state the question of the propensity interpretation of conditional probabilities in a way noticeably different from the way it is usually stated. The acknowledged difference has nothing to do with the mathematics of probability, and has everything to do with the terms in which the question of the propensity interpretation of conditional probabilities is formulated. Indeed, what is at stake is no more whether the interpretation of conditional probabilities that is supposedly implied by the propensity interpretation of absolute probabilities is admissible. It becomes constructing an admissible propensity interpretation of conditionalization.

To that effect, I take into consideration the following property of Bayesian conditionalization: for any probability function Pr and any E such that $\Pr(E) \neq 0$, $\Pr(E|E) = 1$. This property is fundamental to Bayesian conditionalization [6, p. 311], and this is why I consider it rather than any other property of Bayesian conditionalization. Now, this property imposes a constraint on the function c_p constituting the propensity interpretation of conditionalization. Specifically, c_p must be have the following property:

$$\text{For any physical system } S \text{ and any singular event } E \quad (*)$$
$$\text{such the } \Pr_S(E) \neq 0, \Pr_{c_p(S,E)}(E) = 1.$$

Thus, a requirement of minimality on the changes conveyed by conditionalization leads to propose the following definition for c_p:

$$
\begin{aligned}
c_p \quad : \quad &\mathbf{S} \times \mathbf{E} \quad \longrightarrow \quad \mathbf{S} \\
&(S, E) \quad \longmapsto \quad \text{the physical system the most similar to } S \\
&\qquad\qquad\qquad\qquad \text{among those giving probability 1 to } E.
\end{aligned}
$$

In words, the proposition is to interpret conditionalization as the function that associates to the initial physical system, the system from which it differs the least among those that give probability 1 to the conditioning event.

The fact that it makes conditionalization the minimal move required for the satisfaction of Property (*) should count as an asset of the proposed interpretation. Another asset of the proposition concerns inverse conditional probabilities. Humphreys' paradox, it was explained, runs on the question of their evaluation. Now, the proposed interpretation does not give any reason to evaluate them following any of the principles (CI), (FP) or (ZI) that we saw face the paradox under different versions. Let me be a little bit more precise here, and disentangle two arguments. First, let us come back to the photon-emitting system relative to which Humphreys' paradox was originally laid down. Which value does our proposition lead to attribute to the problematic inverse conditional probability $\Pr_{t_1}(I_{t_2}|T_{t_3})$? According to our proposition, this probability is to be interpreted as the probability of I_{t_2} at t_1 relative to the system that is the most similar to the original emitting-photon system among those giving T_3 probability 1. Whatever this system may be exactly, being such that transmission at t_3 has probability 1, it should be such that impingement at t_2 also has probability 1. Thus, our proposition applied to the photon-emitting system arguably leads to a value of $\Pr_{t_1}(I_{t_2}|T_{t_3})$ that could stem from principle (FP). However, this does not mean that our proposition faces the same difficulty as (FP) here. Indeed, as mentioned above, the problem with (FP) arises exactly because it may lead to give value 0 to the conditional probability under consideration — not because it may lead to give it value 1. As a consequence, our proposition is immune to existing versions of Humphreys' paradox for the photon-emitting system. The second point I would like to make here is more general. It essentially consists in claiming that the interpretation I suggest does not seem to lead to any general principle (at all) for the evaluation of inverse conditional probabilities. If this is indeed the case, then it will be difficult to raise against it a difficulty similar to Humphreys' paradox and that would also deal with the evaluation of inverse conditional probabilities.

5 Discussion

Although the proposed interpretation has some assets, it is clear that it also raises objections. Some of them are discussed in this final section. The first objection, presumably, would be to introducing the notion of similarity between physical systems. As an answer, I shall not produce a definition of similarity between physical systems. Rather, I claim that the objection is at least seriously weakened by widening the view. More precisely, it is well-known that the leading approach to counterfactuals is through similarity between possible worlds. Now, it seems to me that accepting similarity between systems is no great deal once similarity between possible worlds has been accepted. Reciprocally, rejecting similarity between physical systems seems to commit to reject similarity between possible worlds — and, along with it, of our best analysis of counterfactuals and an important analysis of causality. As a consequence I do not think that its resorting to similarity between physical systems invalidates the proposed interpretation.

Another difficulty with the proposed interpretation is, precisely, that nothing guarantees that it is indeed an interpretation of conditionalization. It was constructed in such a way that it meets the constraint conveyed by Property ∗. Hence, it accounts for the property of Bayesian conditionalization of being such that for any probability function Pr and any E such that $\Pr(E) \neq 0$, $\Pr(E|E) = 1$. But the proposed interpretation could have been constructed with reference to another property of Bayesian conditionalization. Worse still, I have not given any reason to think that the proposed interpretation indeed accounts for all the properties of Bayesian conditionalization. Once again, I do not have a concluding answer to the objection. Yet I have two arguments to put forward. First, as already noticed, the property on which I focused is fundamental to Bayesian conditionalization. My second argument is comparative. The comparison is no more with alternative interpretations of probability, but with the propensity interpretation of *absolute* probabilities. Actually, there does not exist a conclusive argument to the effect that the propensity theory of absolute probabilities is admissible as an interpretation of the probability calculus — let alone a procedure for measuring propensities. As a consequence, it has to be postulated that propensities behave like absolute probabilities. Now, one can imagine to have an analogous postulate in the conditional case. In other words, merely postulating that the proposed interpretation accounts for the properties of Bayesian conditionalization is a strategy available to a supporter of the propensity interpretation of absolute probabilities.

Acknowledgedly, my arguments against possible objections to the propensity interpretation of conditionalization that I have suggested are rather weak. However, I do not see the content of the proposed interpretation of conditionalization as the most interesting aspect of the present paper. Positively, what I consider as the core contribution of the paper to the debate concerning conditional propensities is precisely its contribution to redefining this debate. I have pointed out that Humphreys, as well as the other participants in the debates consider that an interpretation of conditionalization is contained in Popper's proposition to interpret absolute probabilities as measures of propensities. On the other hand, I have shown that conditionalization requires its own *interpretation* — and I have given an analysis of what it formally is to give such an interpretation. Actually constructing the interpretation, then, is the last job. Maybe the way I have carried it out is not satisfactory; still, others may formulate more convincing propositions. It is my final contention, indeed, that there can be a propensity interpretation of conditional probabilities.

References

[1] Kenny Easwaran. What Conditional Probability Must (Almost) Be, 2005. Unpublished manuscript available on the webpage of the author.

[2] James Fetzer. *Scientific Knowledge: Causation, Explanation, and Corroboration*, volume 69 of *Boston Studies in the Philosophy of Science*. Reidel, 1981.

[3] Alan Hájek. What Conditional Probability Could Not Be. *Synthese*, 137:273–323, 2003.

[4] Paul Humphreys. Why Propensities Cannot Be Probabilities. *Philosophical Review*, 94:557–570, 1985.

[5] Paul Humphreys. Some Considerations on Conditional Chances. *British Journal for the Philosophy of Science*, 55:667–680, 2004.

[6] David Lewis. Probabilities of Conditionals and Conditional Probabilities. *Philosophical Review*, 85:297–315, 1976.

[7] David Lewis. A Subjectivist's Guide to Objective Chance. In Richard Jeffrey, editor, *Studies in Inductive Logic and Probability*, volume II, pages 263–293. University of California Press, 1980.

[8] Peter Milne. Can there Be a Realist Single-case Interpretation of Probability? *Erkenntnis*, 25:129–132, 1986.

[9] Karl Popper. The Propensity Interpretation of the Calculus of Probability, and the Quantum Theory. In Stephan Körner and Maurice H. L. Pryce, editors, *Observation and Interpretation: A Symposium of Philosophers and Physicists. Proceedings of the Ninth Symposium of the Colston Research Society held in the University of Bristol*, pages 65–70. Butterworth's Scientific Publications, 1957.

[10] Karl Popper. The Propensity Interpretation of Probability. *British Journal for the Philosophy of Science*, 10:25–42, 1959.

[11] Frank P. Ramsey. Truth and Probability. In Henry Kyburg and Howard Smokler, editors, *Studies in Subjective Probability*, pages 25–52. Wiley, 1926.

[12] Wesley Salmon. *The Foundations of Scientific Inference*. University of Pittsburgh Press, 1967.

[13] Paul Teller. Conditionalization and Observation. *Synthese*, 26:218–258, 1973.

Benedikt **Löwe**, Eric **Pacuit**, Jan-Willem **Romeijn** (*eds.*)
Foundations of the Formal Sciences VI
Reasoning about Probabilities and Probabilistic Reasoning

Probability: One or Many?

MARIA CARLA GALAVOTTI

Department of Philosophy, University of Bologna, Via Zamboni 38, Bologna, Italy
E-mail: mariacarla.galavotti@unibo.it

1 Foreword

This article deals with the interpretation of probability. There is no doubt that irrespective of its philosophical meaning probability can be applied to all sorts of problems by virtue of its mathematical properties alone. However, it is the philosophical significance of probability that has fostered endless debate. This originates from a long-standing tradition, dating back to the very origin of probability construed in its modern sense, namely as a quantitative notion expressible by means of a function assigning a hypothesis a value ranging in the interval $[0,1]$. Such a tradition holds that probability is not limited to its mathematical properties, but has a broad albeit controversial philosophical scope related to the meaning to be attached to the notion of probability.

More than three and a half centuries after the "official" birth of probability with the work of Blaise Pascal and Pierre Fermat, the dispute on the interpretation of probability is far from being settled. As we shall see, one can distinguish at least four interpretations: frequentism, propensionism, logicism and subjectivism, each of which admits of a number of variants. Upholders of one or the other of these interpretations continue to quibble over the "true" meaning of probability and the "right" method for calculating the probability of simple events, whereas the bulk of the rules for calculating the probabilities of composite events (the probability calculus) meet with a general consensus[1].

Against the tendency to claim that the natural sciences call for a strongly objective, realistic notion of probability, and that the subjective interpreta-

---

[1]Although there are disputes also in this connection, like, e.g., on countable and uncountable additivity.

Received by the editors: 27 July 2007; 19 December 2007.
Accepted for publication: 31 January 2008.

tion is at most applicable to the social sciences, it will be argued that, on closer inspection, subjectivism has the resources to cover all uses of probability, in the natural as well as the social sciences. The following pages will sketch the main interpretations on probability, followed by a discussion of the meaning of probability in the sciences, and finally tackle the question of how to make sense of "objective" probability within the subjective interpretation will be addressed. It will be held that the subjective interpretation has in itself the resources to account for a notion of probability characterised by a robustness that should satisfy the demand of "objectivity" posed by the natural sciences. This conclusion will be substantiated by some ideas borrowed from the writings of Frank Ramsey, Bruno de Finetti and Harold Jeffreys, ideas that have been overlooked by the otherwise extensive literature on these authors.

2   The Interpretations of Probability

Probability has a twofold meaning. On the one hand, it is *empirical* and refers to *chance processes*. Taken in this sense, probability represents a characteristic of phenomena that exhibit a random behaviour. On the other, it is *epistemic*, and taken in this sense it belongs to our knowledge, not to the facts. The dual nature of probability has been convincingly argued for by Ian Hacking in his book *The Emergence of Probability*. As he observes, the two meanings of probability have coexisted for a long time. Through the seventeenth and eighteenth centuries the literature on probability devoted equal attention to the *doctrine of chance* and the *art of conjecture*, built respectively on probability's empirical and epistemic meaning.

At the turn of the nineteenth century Pierre Simon de Laplace imposed the viewpoint that one of such meanings should be privileged over the other and put at the core of the definition of probability. As is well known, Laplace forged the so-called *classical interpretation*, according to which probability is to be calculated as the ratio of the number of favourable cases to that of all possible cases. This is grounded on the assumption that all cases in question are equally possible, lacking information that would lead us to believe otherwise [36].

Laplace's philosophy of probability is rooted in the doctrine of *determinism*, according to which the universe is ruled by the "principle of sufficient reason" stating that all things are brought into existence by a cause. The human mind is incapable of grasping the whole causal network underlying phenomena, but one can conceive of a superior intelligence able to do so. Probability is the tool that enables men to improve their knowledge. Made necessary by the incompleteness of human knowledge, according to this perspective probability is an *epistemic* notion. *It belongs to our knowledge*, rather than being inherent to phenomena.

Laplace's theory was very influential, and dominated the debate on probability for a long time. It can handle a wide array of important applications, but it also faces severe difficulties. Among them the so-called "Bertrand's paradox" that emerges in connection with problems involving an infinite number of possibilities. Moreover, in many situations the set of "equally likely" cases is not determinable. In such situations —think, e.g., of the probability of a biased coin falling on either side, or the probability that a given individual will die within a year— instead of looking for possible cases, one counts the frequency with which events take place, in order to calculate probability. This led to thinking of probability in different terms, giving way to other interpretations of probability.

After Laplace, the debate on the meaning of probability grew wider and wider. One can distinguish four main standpoints, three of which can be classified as *objective*, because they hold that there are correct probability values. They are: (1) *Frequentism*, which qualifies as an *empirical* interpretation; (2) *Propensionism*, which is a *metaphysical* interpretation; and (3) *Logicism*, which is an *epistemic* interpretation. By contrast, the fourth interpretation, namely (4) *Subjectivism*, is *epistemic* and *subjective*. Subjectivism does not share the conviction that there are correct probability values to be univocally determined. Put differently, the first three interpretations entertain in one way or other the idea of *unknown* probability, while subjectivism denies it[2]. These four interpretations are the topics of an ongoing debate, whereas the classical theory is nowadays hardly defendable, due to its commitment to determinism. Obviously this applies to its *philosophical* gist, not to Laplace's method for assessing probabilities, which is still "used in many important applications in which the guiding principles of symmetry are clear and generally well agreed upon" [51, p. 167].

For frequentists, including Robert Leslie Ellis, John Venn, Richard von Mises and Hans Reichenbach, probability is defined as the *limit of the frequency* observed in the initial part of a series of repeatable events, assumed to be indefinitely long. Given that an attribute A has been observed with frequency m/n in the initial part of sequence B, its probability equals $\lim_{n \to \infty} F^n(A, B) = m/n$. This method of evaluating probability, which is at the core of the frequency interpretation, has been called by Hans Reichenbach "rule of induction".

The frequency interpretation is *empirical* and *objective*. It maintains that probability is a *characteristic of phenomena*, which can be empirically analyzed by observing frequencies. Correct —or "true", if one prefers to use this attribute— probability values exist, but are generally unknown. They can be approached by means of frequencies. With Reichenbach, one

---

[2] A survey of the various interpretations of probability is to be found in [24] to which the reader is addressed for a richer exposition of the perspectives sketched here.

could think of probability evaluation as a *self-correcting procedure*, by which repeated applications of the rule of induction lead to results that in the long run will converge towards the correct value [46, 47].

Being defined and calculated with reference to sequences of repeatable events, according to frequentism probability cannot cover single occurrences of events. In other words, it makes no sense to refer to the probability of single events. This creates a *single case problem* that is the most puzzling aspect of the frequency interpretation.

It is noteworthy that unlike the classical interpretation, frequentism is fully compatible with *indeterminism*. As a matter of fact, Richard von Mises regards this feature as the main advantage of frequentism over Laplace's view [52].

Anticipated by Charles Sanders Peirce, the propensity theory was put forward in the late fifties by Karl Popper to solve the problem of interpreting single case probability attributions, especially those occurring in quantum mechanics. According to the propensity theory, probability is a *non-observable property "of the generating conditions"* [42, p. 34] surrounding the occurrence of non-deterministic events. In an attempt to give them a strongly *objective* character, Popper qualifies propensities as *physically real*. This obviously involves a commitment to indeterminism.

While the propensity theory was initially conceived by Popper as a variant of the frequency interpretation, in the eighties he made it the focus of a wider program meant to account for all sorts of causal tendencies operating in the world [44]. He then saw propensities as "weighted possibilities", or expressions of the tendency of a given experimental set-up to realize itself upon repetition, emphasizing single experimental arrangements rather than sequences of generating conditions. In so doing, he laid down the so-called "single-case propensity interpretation". At that point he described propensity as "a new physical (or perhaps metaphysical) hypothesis" [43, p. 360] analogous to Newton's "force". At the core of Popper's theory lies a distinction between *probability statements*, or statements about *propensities*, which are about frequencies in *virtual* sequences of experiments, and *statistical statements*, which express relative frequencies observed in *actual* sequences of experiments. Being based on frequencies that have actually been observed, statistical statements can be used to *test* propensity statements. In spite of their metaphysical character, the latter are therefore testable against observable frequencies, and for this reason the propensity theory is qualified as *empirical*.

In the debate that followed, the single-case propensity interpretation became widely popular among philosophers of science, but also another variant, called "long run propensity interpretation", took shape. Its supporters include Donald Gillies, on whose work something will be added later.

The Laplacean tradition of epistemic probability branched into two different approaches: logicism and subjectivism. According to the logical interpretation, the theory of probability belongs to logic, and probability is a *logical relation* between propositions, one of which describes a body of evidence while the other states a hypothesis. A natural development of the idea that probability is epistemic and pertains to our knowledge of facts rather than to the facts themselves, this approach, in comparison with Laplace's, places more emphasis on the logical aspect of probability, which is meant to give it an analytical and objective foundation.

Anticipated by Leibniz, the logical interpretation was embraced by a number of authors including Bernard Bolzano, Augustus De Morgan, George Boole, William Stanley Jevons, John Maynard Keynes, William Ernest Johnson, Harold Jeffreys, Ludwig Wittgenstein, Friedrich Waismann and Rudolf Carnap, who developed its most sophisticated version. All of these authors share a view of probability as epistemic and *objective* and regard probabilities as providing objective grounds for belief. Stress is put on *rationality* and on the idea that beliefs based on logical probabilities are rational. Rationality is grounded on the *objective character* of logical probability, which requires that in the light of the same evidence there is only one correct (rational) probability assignment.

This conviction is not shared by the *subjective* interpretation of probability of Frank Plumpton Ramsey and Bruno de Finetti. According to subjectivism, probability is the degree of belief actually entertained by someone in a state of uncertainty regarding the occurrence of an event. Probability as degree of belief is a primitive notion, which requires an *operative definition* specifying a way of measuring it. Various options to achieve this goal are available, like the method of bets, dating back to the seventeenth century, according to which one's degree of belief in the occurrence of an event can be expressed by means of the odds at which one would be ready to bet. Other operative definitions have been devised, including Ramsey's method based on *preferences*, and *penalty methods* adopted by de Finetti in a number of writings.

The cornerstone of the subjective interpretation is the notion of *coherence*. Applied to betting ratios, coherence guarantees avoidance of a sure loss. Since it can be shown that coherent degrees of belief satisfy the laws of probability, coherence can be taken as the only condition of acceptability that needs to be imposed on degrees of belief. In other words, for subjectivists once degrees of belief are coherent, there is no further demand of rationality to be met: *all coherent probability functions are admissible.*

Unlike logicists, subjectivists believe that probability evaluations are not univocally determined by evidence. There are no correct probability values, nor unknown probabilities: it is admitted that two people facing the same body of evidence give different evaluations of probability. The other cornerstone of subjectivism, namely the result known as "de Finetti's representation theorem", shows that the adoption of Bayes' method taken in conjunction with exchangeability leads to a convergence between degrees of belief and frequencies. For the subjectivist de Finetti this makes "objective" probability, namely the idea that probability should be uniquely determined, a useless notion. Instead, one should be aware that probability evaluations depend on both subjective and objective elements, and refine probability appraisers by means of calibration methods.

3   Probability in the Sciences

Since von Mises, frequentism has become the "official" interpretation of probability in science. In the last chapter of *Probability, Statistics and Truth* von Mises identifies the typical areas that can be treated probabilistically as: the kinetic theory of gases; Brownian motion; radioactivity; Planck's theory of black-body radiation; and claims to be confident that frequentism can be extended without major problems to quantum mechanics [53, p. 211]. But the frequency theory *does* have a problem in that connection: it cannot be applied to the single case, while quantum mechanics regularly talks about single case probabilities, like the probability that a certain atom is in a given state.

Reichenbach made an attempt to make sense of single case probabilities with his theory of *weight*. The idea here is that a probability evaluation — Reichenbach calls it a *posit*— regarding a single occurrence of an event receives a weight from the probabilities attached to the reference class to which the event in question has been assigned, which must obey a criterion of *homogeneity*. This is to say that the reference class should include as many cases as possible similar to the one under consideration, while excluding dissimilar ones. Similarity is to be taken relative to the properties that are considered relevant, and homogeneity is obtained through successive partitions of the reference class by means of statistically relevant properties. A reference class that cannot be further partitioned in this way is *homogeneous*. For instance, if one wanted to assign a weight to the probability that a given individual will die within the next five years, the reference class ought to be chosen on the basis of a series of properties taken to be relevant, like age, sex, occupation, nationality, health status, and so on, whereas irrelevant properties, like hair colour, or shoe size, should be excluded. The probability of death of individuals belonging to the reference class so obtained, determinable on the basis of observed frequencies, will give the weight to be assigned to the survival of a given individual. In Reichenbach's words: "A weight is what a degree of probability becomes if it is applied to a single case"[47, p. 314].

However ingenious, Reichenbach's solution to the single case problem is not free from difficulties, because one can never be absolutely sure that all the properties that are relevant to some phenomenon have been taken into account. Identifying the proper reference class is obviously a delicate matter, which poses serious problems, well-known and widely discussed by statisticians. The homogeneity requirement is therefore an unfulfilled ideal, which can at best have a heuristic import.

The single case problem is also at the origin of Popper's propensity theory, meant to account for quantum mechanical probabilities. No doubt the propensity theory allows us to talk about single case probabilities, but it also gives rise to serious problems. One of them is posed by *Humphreys'* argument pointing to the *impossibility of interpreting inverse probability* in propensionist terms. If Humphreys is correct, the dispositional character of propensities, defined as tendencies to produce certain outcomes, confers on them a peculiar asymmetry, which clashes with the symmetry characterizing probability as described by the probability calculus. Most notably, such asymmetry makes Bayes' theorem inapplicable to propensities [30]. Humphreys' problem provoked an ongoing debate, which registers a wide array of different positions. Among those who take it seriously are Wesley Salmon, who embraces a *causal interpretation of propensities* [50] according to which propensities are causal tendencies rather than probabilities, and Donald Gillies, who instead establishes a stronger link between propensities and frequencies, embracing a *long-run propensity interpretation* that takes propensities as tendencies to produce long-run frequencies [25]. Unlike these authors, David Miller believes that Humphreys' problem does not arise within his own single-case propensity interpretation, which makes propensities dependent on the *complete state of the universe* at a given time[3]. Needless to say, this is a strong requirement, hardly ever fulfilled.

Furthermore, the propensity theory seems to be affected by the *reference class problem* no less than frequentism. In fact, in order to calculate the value of propensities one has to depend on frequencies. But obviously such frequencies will have to be referred to a reference class. Far from being removed, the problems encountered by frequentism in this connection are only displaced.

So much for the empirical interpretations of probability. Insofar as epistemic probability is concerned, the subjective interpretation has been considered by many authors suitable for the social sciences, but unsuitable for the natural sciences. Subjectivism has never been taken too seriously by physicists, who are usually more sympathetic to the frequency interpretation. Physicists' aversion to subjectivism might have been prompted by de Finetti's uncompromising attitude, that led him to reject notions like

---

[3]Cf. [40] and also [41] for further discussion.

"chance", "objective probability" and "randomness"[4]. Under the spell of
von Mises' work, even the fathers of quantum mechanics, including Werner
Heisenberg and Max Born, showed confidence in the powers of frequentism.

   Notable exceptions to the frequentist tendency are the physicists Richard
T. Cox, Edwin T. Jaynes and Harold Jeffreys, upholders of a form of
Bayesianism that goes hand in hand with logical probability. The result
known as "Cox's theorem" derives the laws of probability from a set of pos-
tulates which are meant as fixing "plausibility" conditions[5]. Building on
Cox's result, which he combined with information theory, and more specifically with the idea that maximum entropy can prove a useful criterion to
be applied to statistical inference, Jaynes reformulated Bayesian method as
the "logic of science" [49, 2] and worked on its application to a wide array
of problems.

   The geophysicist Harold Jeffreys —to whom Jaynes' book *Probability
Theory: The Logic of Science* is dedicated— devoted great attention to
the philosophical aspects of probability and developed an original epistemology[6]. While embracing an epistemic notion of probability, Jeffreys regarded subjectivism as a theory of expectation, not suited to interpreting
probability in science. Convinced that science calls for a "pure" notion of
probability, free from reference to mathematical expectation, Jeffreys embraced a combination of Bayesianism with the *logical* view of probability
as *reasonable* belief, corresponding to the degree of belief warranted by a
certain body of evidence, which is uniquely determined [32]. However —as
we will see in Section 4— he put forward a view of chance and objective
probability that fits in with subjectivism.

   There is no doubt that to some the subjective interpretation appears
somewhat arbitrary. A case in point is that of forensic scientists, who tend
to discard the subjective slant and turn to logical probability, in an attempt to safeguard objectivity (*cf.*, e.g., [37]). In so doing, in spite of the
fact that they admittedly deal with opinions, they are presumably reassured
by the promise of objectivity conveyed by the term "logical". By contrast,
the idea that subjectivism is a theory of expectation, appropriate for those
settings requiring a decision on the best course of action, makes it palat-
able to economists and econometricians, who are habitually concerned with
personal opinions and expectations.

   With few exceptions, the logical interpretation has never become popular
with scientists. Carnap's writings, which convey a formally sophisticated
version of logicism [3, 4], are almost unknown outside the restricted circle

---

[4]This is argued at length in [19]. De Finetti's refusal of such notions is rooted in his
anti-realistic philosophy, described in [17].

[5]Cf. [6, 7]. On Cox's theorem, which has provoked extensive discussion, cf. [27, 5].

[6]Jeffreys' epistemology is discussed in some detail in [22] and [24].

of logicians and philosophers of science. Cox, Jaynes and Jeffreys met with
more favour within scientific circles, and recently attracted the attention
of philosophers of science and computer scientists. But it should be added
that literature more often associates the work of these authors with the
foundations of Bayesianism than with the interpretation of probability: two
issues that in our opinion should not be made to overlap. Furthermore,
insofar as Jeffreys and Jaynes are concerned, the literature has focussed
especially on their attempts to suggest "objective" criteria for the choice of
prior probabilities, which for Jaynes should be guided by the "principle of
maximum entropy", and for Jeffreys should satisfy a "simplicity postulate"[7]
Jeffreys' work, by the way, seems to count more followers among statisticians
and economists than among physicists.

Faced with such a diversified picture, some authors quit the search for
a univocal interpretation of probability apt to cover all applications, to
adopt a pluralistic perspective. Among them Donald Gillies, who in *Philosophical Theories of Probability* [25] embraces a *pluralistic* approach to the
interpretation of probability. According to Gillies the propensionist interpretation —taken in its long-run version— is apt to interpret probabilities
belonging to the natural sciences, which require an objective notion of probability. By contrast the social sciences, that do not deal with independent
repetitive events but rather with human agents and their beliefs, call for
the subjective interpretation of probability.

4 Subjectivism, Objectivism and Objectivity

Let us next address the question: Can one make sense of "objective probability" within the subjective interpretation? Bruno de Finetti maintained
that *"probability does not exist"*, and wanted this sentence printed in capi-
tal letters in the preface to the English edition of his *Theory of Probability*
[13]. As already observed, his insistence upon this claim is likely to have
been an obstacle to the diffusion of the subjective interpretation beyond
those contexts in which probability is taken as a theory of expectation and
a basis for decision-making. However, it should be added that de Finetti's
claim should not be taken to mean that probability can be assigned whatever value, or that there is no problem of objectivity in connection with the
evaluation of probability. This is not what de Finetti meant. De Finetti
did not deny that there is a problem of objectivity of the evaluations of
probability, he rather opposed the distortion of "identifying objectivity and
objectivism", deemed a "dangerous mirage" [11, p. 344]. De Finetti does
not fight against *objectivity*, but against *objectivism* taken as the idea that
probability depends entirely on some aspects of reality and is to be uniquely
determined by evidence.

---

[7]A discussion of the simplicity postulate is to be found in [28].

The cornerstone of de Finetti's position is the conviction that exchangeability taken in combination with Bayes' rule makes the notion of "objective", or "unknown" probability idle. Exchangeability is for de Finetti the tool that allows us to express correctly the idea that is usually conveyed by the phrase "independent events with constant but unknown probability". If we take an urn of unknown composition, relative to each of all possible compositions of the urn drawings can be seen as independent with constant probability. But "...what is unknown is the composition of the urn not the probability: the latter is always known and depends on the subjective opinion about the composition, which opinion is modified as new drawings are made, and observed frequencies are taken into account" [16, p. 163]. Used in connection with Bayes' rule, exchangeability allows subjective probability judgments to be updated by taking into account observed frequencies. While bridging the gap between degrees of belief and observed frequencies, this method performs what de Finetti used to call the "reduction" of objective to subjective probability.

Albeit for de Finetti objective probability (uniquely determined, inherent to facts) does not exist, there is a *problem of objectivity*, which in subjectivistic terms can be rephrased as the problem of devising good probability appraisers. To be sure, the evaluation of probability should take into account all available evidence, including frequencies and symmetries. However, it would be a mistake to put these elements, which are useful ingredients of the *evaluation* of probability, at the basis of the *definition* of probability. This is the mistake imputed to other interpretations, that define probability in terms of frequency or symmetry, and on this account de Finetti sharply criticizes frequentism, logicism and the classical approach. Precisely as one should not confuse (the refusal of) objectivism with (the problem of) objectivity, one should not conflate the definition and the evaluation of probability.

According to de Finetti the evaluation of probability results from "the conjunction of both objective and subjective elements at our disposal" [12, p. 366]. Both components, "(1) the objective component, consisting of the evidence of known data and facts; and (2) the subjective component, consisting of the opinion concerning unknown facts based on known evidence" [14, p. 7], are essential, and the explicit recognition of the role played by subjective elements within probability judgments is a prerequisite for the appraisal of objective elements. Far from being the product of ready-made recipes applicable to all situations, the evaluation of probability is a context-dependent procedure, because both its objective and subjective components depend in various respects on the context. Factual evidence must be collected carefully and skillfully, its exploitation depending on the judgment on what elements are relevant to the problem under consideration and enter into the evaluation of probabilities. Moreover, the collection and exploitation of evidence depends on considerations that vary according to the context (often these are economic considerations, but other elements could be relevant too). Equally subjective for de Finetti is the decision, largely dependent on personal experience and individual judgement, on how to let belief be influenced by objective elements.

To sum up, though the intrinsic context sensitivity of probability evaluation makes the idea of *absolute objectivity* nonsensical, a problem of objectivity still arises, and is addressed by de Finetti. At the core of a viable notion of objectivity he puts the possibility of evaluating probability assessments by means of *calibration*. His approach in this connection is based on penalty methods, like the so-called "Brier's rule", named after the meteorologist Brier who applied it to weather forecasts. Methods like scoring rules are perfectly legitimate and justifiable within a subjectivist framework. In de Finetti's words: "...though maintaining the subjectivist idea that no fact can prove or disprove belief, I find no difficulty in admitting that any form of comparison between probability evaluations (of myself, of other people) and actual events may be an element influencing my further judgment, of the same status as any other kind of information" [10, p. 360].

De Finetti is led by his anti-realist philosophy to refuse, together with objective and unknown probability, the notions of *chance* and *physical probability*. These are taken as metaphysical entities, having no right of citizenship in the realm of subjectivism. However, in the posthumous volume *Filosofia della probabilità* one finds the admission that those evaluations of probability that are grounded on scientific theories have a stronger character than those encountered in everyday life. For instance, referring to the probability distributions belonging to statistical mechanics, de Finetti claims that "they provide more solid grounds for subjective opinions" [16, p. 52].This allows for the conjecture that late in his life he entertained the idea that probabilities encountered in science derive a peculiar "robustness" from scientific theories[8].

A similar idea inspires Frank Ramsey's notions of chance and probability in physics, which are accounted for within the framework of subjective probability. This is done by referring such notions to *systems of beliefs* that typically include widely accepted generalisations. For Ramsey "Chances are degrees of belief within a certain system of beliefs and degrees of belief; not those of any actual person, but in a simplified system to which those of actual people, especially the speaker, in part approximate". Chances can be seen as *objective* "in that everyone agrees about them, as opposed, e.g., to odds on horses" [45, pp. 104–105].

---

[8]For more details, cf. [21, 23].

Ramsey also defines *probability in physics* as chance referred to a more complex system, namely to a system making reference to scientific theories. In other words, probabilities occurring in physics are derived from physical theories. They are *ultimate chances*, in the sense that within the theoretical framework in which they occur there is no way of replacing them with deterministic laws. Their objective character derives from the objectivity characterizing theories. In turn, the objective character usually ascribed to theories is warranted by acceptance on the part of the scientific community: only those theories which gain *universal assent* in the long run are accepted and taken as true. In similar terms Ramsey speaks of a "true scientific system", referring to a system to which the opinion of everyone, grounded on experimental evidence, will eventually converge. Summing up, for Ramsey chances are theoretical constructs, but they do not express realistic properties of "physical objects", whatever meaning is attached to this expression. Chance attributions indicate a way in which beliefs in various facts belonging to science are guided by scientific theories[9].

Being inspired by a similar (pragmatical) philosophy of probability and grounded on the same (subjective) interpretation, Ramsey's notion of chance could be seen as the natural complement to de Finetti's position. Taken together, their views provide the ingredients of a version of subjectivism apt to account for probability in science.

Additional hints are offered by Harold Jeffreys, who regards chance as a *limiting case of everyday assignments*. Chance occurs in those situations in which "given certain parameters, the probability of an event is the same at every trial, no matter what may have happened at previous trials" [34, p. 46]. For instance, chance "will apply to the throw of a coin or a die that we previously know to be unbiased, but not if we are throwing it with the object of determining the degree of bias. It will apply to measurements when we know the true value and the law of error already... It is not numerically assessable except when we know so much about the system already that we need to know no more" [31, p. 329].

Jeffreys also discusses *probability in physics*, calling attention to those fields where "some scientific laws may contain an element of probability that is intrinsic to the system and has nothing to do with our knowledge of it" [33, p. 284]. This is the case with quantum mechanics, whose account of phenomena, at least according to the Copenhagen interpretation, is irreducibly probabilistic. Unlike the probability (chance) that a fair coin falls heads,*intrinsic* probabilities do not belong to our description of phenomena, but to the *theory itself*. In Jeffreys' words: "whether there is or not" intrinsic probability, "it can be discussed in the language of epistemological probability" [33, p. 284]. In other words, Jeffreys is convinced that epis-

---

[9]For more details on Ramsey's view of chance and probability in physics, cf. [18, 20].

temic probability can do the whole job and there is no need to admit of other notions of probability. In fact, he criticizes Carnap for admitting two notions of probability: probability$_1$, or logical probability, and probability$_2$, or probability as frequency.

A very interesting aspect of Jeffreys' work lies in his constructivist approach to the notions of *objectivity* and *reality*. For him what is "objective" or "real" is established by inference from experience through the adoption of statistical methodology [32, p. 336]. Like the concept of chance, that of objectivity acquires a definite meaning through scientific methodology. It is only after the rules of induction "have compared it with experience and attached a high probability to it as a result of that comparison" that a general proposition can become a law. In this procedure lies "the only scientifically useful meaning of 'objectivity' " [32, p. 336]. In spite of his logicism, Jeffreys' ideas in that connection are very much in tune with subjectivism.

5   Concluding Remarks

Ramsey's work points the way to a version of subjectivism that can accommodate the notion of "objective" probability. Similarly, Jeffreys suggests how notions like chance, probability in physics and objectivity can be addressed within the epistemic framework. These ideas of Ramsey and Jeffreys can fill the gap left open by de Finetti's refusal of objective probability. This constitutes an important step towards building a form of subjectivism flexible enough to cover all applications of probability. Another important ingredient of such a perspective is the distinction made by de Finetti between the evaluation and the definition of probability. Granted that probability is the expression of a degree of belief, when it comes to the evaluation of probability it is admitted that objective elements play an essential role. They are combined with subjective elements to form probability judgements. In some cases the objective component is preponderant, as it happens when probability values are derived from scientific theories. If the problem is addressed in these terms, probability can acquire an objective meaning also within the subjective perspective, as it was clearly seen by Ramsey.

Incidentally, the kind of subjectivism envisaged here also lends itself to application by forensic science. A plea in the same direction is advanced by Philip Dawid, who observes: "The subjectivist philosophy is not to say that anything goes: in the light of whatever relevant evidence may be available, certain opinions will be more reasonable than others. This seems to correspond to the legal conceptions of the 'reasonable man' and 'reasonable doubt'. In simple statistical problems, it can be shown that differing initial subjective distributions will be brought into ever closer and closer agreement when updated through the incorporation of sufficiently extensive observational evidence. In legal applications the conditions for this convergence will often not apply, but even so there may be certain probabilistic ingredients and conclusions that can be regarded as reasonable by all reasonable parties" [9].

In recent years, considerable effort has been devoted to the notion of objective probability, in the conviction that this is required by natural science. Important work in this direction has been done by authors operating under the label "objective Bayesianism". Among them Jon Williamson, who contemplates an "ultimate belief notion of chance" [54] which dovetails the version of subjectivism heralded here. As a matter of fact, the authors involved in the "objective Bayesianism" project do not openly embrace subjectivism, and prefer to speak of a "Bayesian interpretation of probability".

This locution is becoming increasingly popular. It is also adopted, among others, by Colin Howson, a strenuous defender of Bayesianism as *the* logic of science. Appealing to the work of Cox and I. J. Good (*cf.* especially [26]), Howson brings both deductive and probabilistic reasoning under the umbrella of a "general theory of consistency". In this framework "deductive logic is the logic of consistent assignments of truth values, subject to the usual classical truth-definition constraints, while the logic of uncertainty is that of consistent assignments of probability" [29, p. 41].

According to the present writer, Bayesianism and the interpretation of probability represent distinct, albeit interconnected, issues. Bayesianism is a theory of inference, which *can*, and *is* shared by upholders of different interpretations of probability, namely subjectivists, but also logicists (e.g., Jaynes and Jeffreys) and frequentists (e.g., Hans Reichenbach).

To the question: "Probability: one or many?" Bruno de Finetti would answer, echoing Luigi Pirandello, that probability is "one, none, a thousand". His pluralism, however, does not concern the definition of probability, but rather its evaluation. This is not to say that subjectivism is an "anything goes" affair. Even less does it mean that subjective probability only concerns everyday life, because subjectivism also has the resources to account for all uses of probability.

References

[1] Terence Anderson, David Schum, and William Twining. *Analysis of Evidence. Law in Context.* Cambridge University Press, 2nd edition, 2005.

[2] Larry Bretthorst, editor. *E. T. Jaynes: Probability Theory: The Logic of Science.* Cambridge University Press, 2003.

[3] Rudolf Carnap. *Logical Foundations of Probability*. University of Chicago Press, 1950. 2nd edition 1962.

[4] Rudolf Carnap. The Aim of Inductive Logic. In Ernest Nagel, Patrick Suppes, and Alfred Tarski, editors, *Logic, Methodology and Philosophy of Science*, pages 303–318, 1962. Reprinted in [38, pp. 104–20].

[5] Mark Colyvan. The Philosophical Significance of Cox's Theorem. *International Journal of Approximate Reasoning*, 37:71–85, 2004.

[6] Richard T. Cox. Probability, Frequency, and Reasonable Expectation. *American Journal of Physics*, 14:1–13, 1946.

[7] Richard T. Cox. *The Algebra of Probable Inference*. Johns Hopkins University Press, 1961.

[8] Andrew Dale, editor. *Pierre-Simon Laplace. Philosophical Essay on Probabilities*. Springer, 1995.

[9] Philip Dawid. Probability and Proof, 2005. Appendix I to [1]. Available online on the Cambridge University webpage of the book.

[10] Bruno de Finetti. Does it Make Sense to Speak of 'Good Probability Appraisers'? In Irvin J. Good, Alan J. Mayne, and Maynard Smith, editors, *The Scientist Speculates. An Anthology of Partly-Baked Ideas*, pages 357–364. Basic Books, 1962.

[11] Bruno de Finetti. Obiettività e oggettività: critica a un miraggio. *La Rivista Trimestrale*, 1:343–367, 1962.

[12] Bruno de Finetti. Bayesianism: Its Unifying Role for Both the Foundations and the Applications of Statistics. *Bulletin of the International Statistical Institute: Proceedings of the 39th Session*, 39:349–368, 1973.

[13] Bruno de Finetti. *Theory of Probability*. Wiley, 1974.

[14] Bruno de Finetti. The Value of Studying Subjective Evaluations of Probability. In Carl-A. Staël von Holstein, editor, *The Concept of Probability in Psychological Experiments*, pages 1–14. Reidel, 1974.

[15] Bruno de Finetti. *Filosofia della probabilità*. Il Saggiatore, 1995. English translation in [16].

[16] Bruno de Finetti. *Philosophical Lectures on Probability*. Springer, 2008. English translation of [15].

[17] Maria Carla Galavotti. Anti-realism in the Philosophy of Probability: Bruno de Finetti's Subjectivism. *Erkenntnis*, 31:239–261, 1989.

[18] Maria Carla Galavotti. F.P. Ramsey and the Notion of 'Chance'. In Jaakko Hintikka and Klaus Puhl, editors, *The British Tradition in XXth Century Philosophy*, pages 330–340. Hölder-Pichler-Tempsky, 1995.

[19] Maria Carla Galavotti. Operationism, Probability and Quantum Mechanics. *Foundations of Science*, 1:99–118, 1995.

[20] Maria Carla Galavotti. Some Remarks on Objective Chance F.P. Ramsey, K.R. Popper and N.R. Campbell. In Maria L. Dalla Chiara, Roberto Giuntini, and Federico Laudisa, editors, *Language, Quantum, Music*, pages 73–82. Kluwer, 1999.

[21] Maria Carla Galavotti. Subjectivism, Objectivism and Objectivity in Bruno de Finetti's Bayesianism. In David Corfield and Jon Williamson, editors, *Foundations of Bayesianism*, pages 161–174. Kluwer, 2001.

[22] Maria Carla Galavotti. Harold Jeffreys' Probabilistic Epistemology: Between Logicism and Subjectivism. *British Journal for the Philosophy of Science*, 54:43–57, 2003.

[23] Maria Carla Galavotti. Kinds of Probabilism. In Paolo Parrini, Wesley C. Salmon, and Merrilee H. Salmon, editors, *Logical Empiricism: Historical and Contemporary Perspectives*, pages 281–303. University of Pittsburgh Press, 2003.

[24] Maria Carla Galavotti. *Philosophical Introduction to Probability*. CSLI, 2005.

[25] Donald Gillies. *Philosophical Theories of Probability*. Routledge, 2000.

[26] Isidore Jacob Good. *Probability and the Weighing of Evidence*. Griffin, 1950.

[27] Joseph Halpern. Technical Addendum. Cox's Theorem Revisited. *Journal of AI Research*, 11:429–435, 1999.

[28] Colin Howson. On the Consistency of Jeffreys's Simplicity Postulate, and its Role in Bayesian Inference. *Philosophical Quarterly*, 38:68–83, 1988.

[29] Colin Howson. Scientific Reasoning and the Bayesian Interpretation of Probability. In Wenceslao J. González and Jesús Alcolea, editors, *Contemporary Perspectives in Philosophy and Methodology of Science*, pages 31–45. Netbiblo, 2006.

[30] Paul Humphreys. Why Propensities Cannot Be Probabilities. *Philosophical Review*, 94:557–570, 1985.

[31] Harold Jeffreys. The Problem of Inference. *Mind*, 45:324–333, 1936.

[32] Harold Jeffreys. *Theory of Probability*. Clarendon Press, 1939. 2nd edition with modifications 1948.

[33] Harold Jeffreys. The present Position in Probability Theory. *British Journal for the Philosophy of Science*, 5:275–289, 1955. Also in [35, pp. 421–435].

[34] Harold Jeffreys. *Scientific Inference*. Cambridge University Press, 2nd edition, 1957. 1st edition 1931.

[35] Harold Jeffreys and Bertha Swirles, editors. *Collected Papers of Sir Harold Jeffreys on Geophysics and Other Sciences*, volume VI. Gordon and Breach Science Publishers, 1971–1977.

[36] Pierre-Simon Laplace. *Essai philosophique sur les probabilités*. Courcier, 1814. English edition in [8].

[37] Larry Laudan. *Truth, Error and Criminal Law: an Essay in Legal Epistemology*. Cambridge University Press, 2006.

[38] Sidney A. Luckenbach, editor. *Probabilities, Problems, and Paradoxes*. Dickenson, 1972.

[39] David Mellor, editor. *Frank P. Ramsey. Philosophical Papers*. Cambridge University Press, 1990.

[40] David Miller. *Critical Rationalism: A Restatement and Defense*. Open Court, 1994.

[41] David Miller. Propensities May Satisfy Bayes's Theorem. In Richard Swinburne, editor, *Bayes's Theorem*, pages 111–116. Oxford University Press, 2002.

[42] Karl Popper. The Propensity Interpretation of Probability. *British Journal for the Philosophy of Science*, 10:25–42, 1959.

[43] Karl Popper. *Realism and the Aim of Science*. Hutchinson, 1983.

[44] Karl Popper. *A World of Propensities*. Thoemmes Press, 1990.

[45] Frank P. Ramsey. Truth and Probability. In Richard B. Braithwaite, editor, *Frank P. Ramsey, The Foundations of Mathematics and other Logical Essays*, pages 156–198. Kegan, Paul, Trench, Trubner & Co., 1931. Reprinted in [39, pp. 52–109].

[46] Hans Reichenbach. *Wahrscheinlichkeitslehre*. Sijthoff, 1935. English expanded version in [48].

[47] Hans Reichenbach. *Experience and Prediction*. University of Chicago Press, 1938.

[48] Hans Reichenbach. *The Theory of Probability*. University of California Press, 2nd edition, 1971.

[49] Roger Rosenkrantz, editor. *E. T. Jaynes: Papers on Probability, Statistics and Statistical Physics*, volume 158 of *Synthese Library*. Reidel, 1983.

[50] Wesley Salmon. Propensities: A Discussion Review. *Erkenntnis*, 14:183–216, 1979.

[51] Patrick Suppes. *Representation and Invariance of Scientific Structures*. CSLI, 2002.

[52] Richard von Mises. *Wahrscheinlichkeit, Statistik und Wahrheit*. Springer, 1928.

[53] Richard von Mises. *Probability, Statistics and Truth*. Dover Publications, 1957. Reprint of the 1939 Allen and Unwin edition. English translation of [52].

[54] Jon Williamson. Philosophies of Probability: Objective Bayesianism and its Challenges. In Andrew Irvine, editor, *Handbook of the Philosophy of Mathematics*. Elsevier, 2004. Volume 9 of *Handbook of the Philosophy of Science*.

Benedikt **Löwe**, Eric **Pacuit**, Jan-Willem **Romeijn** (*eds.*)
Foundations of the Formal Sciences VI
Reasoning about Probabilities and Probabilistic Reasoning

Probability: from doctrine to theory
Emergence of early modern probability calculus

ANNE-SOPHIE GODFROY-GENIN

Université Paris 12 & École Normale Supérieure de Cachan, 61 avenue du Président Wilson, 94235 Cachan, France
E-mail: anne-sophie.godfroy@ens-cachan.fr

1 Introduction

How did the calculus of probabilities emerge? This issue has been discussed in numerous articles and books [8, 9, 10, 13, 16, 18, 22, 23, 24, 25, 26, 27, 28, 29, 32]. This paper will focus on the role the Port-Royal Logic [1, 2] played in the emergence of probabilistic calculus. This book is seldom mentioned by probability historians who privilege chances' calculus (Pascal, Huygens), statistical approach (mortality tables)[1] or early probabilistic thinking before Pascal (e.g., [7, 13]). A noteworthy exception is Hacking's book [16] that dedicates a full chapter to the Port-Royal Logic. However, we will show how this book played a major role in the conceptual construction of the notion of probability bridging the gap between the philosophical concept of probability as inherited from Aristotle — though unstable and often modified, the calculus of chances explored by Pascal and Huygens and early pre-statistical ideas about regularity of natural and social phenomena such as the occurrence of common accidents like being hurt by thunder. The Port-Royal Logic was almost certainly an essential step between the "doctrine of probability" or "doctrine of probabilism" as expressed by the Jesuit casuists[2], and what has been called "theory of probability". The main authors who conceived this "doctrine" are Bartolomeo de Medina, Vasquez, Suarez and Caramuel. By "theory of probability", we mean the mathematic calculus of probability as it has been theorised at the beginning of the XVIIth century by Jakob Bernoulli first, then by Abraham de Moivre, Pierre

---

[1]E.g., the recent book [4] in French confirms this trend.
[2]Deman in the article [12] provides a very useful description of this doctrine and its evolution.

Received by the editors: 3 November 2007; 9 October 2008; 7 November 2008.
Accepted for publication: 25 March 2009.

Raymond de Montmort and followers. Laplace [5] as well as more recent historians as Todhunter explicitly refer to their works when speaking about the "theory of probability", a term now in the common language.

The legend of probability history, as expressed by Poisson [20], says: *Un problème relatif aux jeux de hasard, proposé à un austère janséniste par un homme du monde, a été à l'origine du calcul des probabilités.* In other words, Chevalier de Méré's questions about dice games would have inspired Pascal, who is supposed to have discovered probability calculus. The idea is appealing and was reproduced in many subsequent books[3]. However "probability" in Pascal's works always means the Jesuit doctrine of probability, as criticized in the *Lettres Provinciales*. "Probability" has nothing to do with geometry of chance and the wager. Therefore, it becomes necessary to investigate a much more complex history to understand how ideas have been woven to provide the conceptual framework of probability calculus.

When we try to have a closer look, the mixed origins of probability calculus becomes obvious and we discover that it stands at the crossroads of at least three traditions:

First, calculations on games of chance with possible applications to business and insurance [9, 26];

Second, calculations on large collections of data, such as mortality tables, to identify regular phenomena and eventually applications of those ideas to law through probability of true and false testimonies according to the internal or external contexts;

Third, philosophical and theological concept of qualitative probability inherited from Aristotle and Aquinas, and revisited by Jesuit casuists during the XVIth and XVIIth centuries [7, 13, 25].

In 1654, Pascal's and Fermat's solutions[4] of the division problem allowed a new idea: quantifying uncertainty, but we are still far from a probability calculus. Not only the word itself is missing, but also the idea. Pascal and Fermat, as Huygens, worked on the idea of expectation, loss chances and gain chances, but never on probability. For Pascal this is amazing given the many pages devoted to probability (in another meaning) in the *Lettres Provinciales*. On the contrary, modern translations or hasty comments often pretend that "probability" appears in the division problem solution,

---

[3]*Cf.* Todhunter's [35] which has been a reference on this topic for many years, or Philippe Sellier, in an introduction to Pascal's *Pensées*: "*Sous l'influence de Méré, adonné au jeu, [Pascal] jette les bases du calcul des probabilités et compose le Traité du Triangle arithmétique.* [31, p. 96]"

[4]See letters between Pascal and Fermat in summer 1654, and the Latin and French versions of the *Traité du Triangle arithmétique*.

which is obviously not the case. Nevertheless, as early as 1662,[5] the authors of the Port-Royal Logic connected Pascal's wager to the last chapters about human belief, which mentions both Pascal and the word "probability" (Part 4, Chapter 13). This fact appears as a clear sign that the meaning of probability has changed in the years immediately following 1654, and that, even if Pascal was not directly responsible for it, Pascal's work reinterpreted in the Port-Royal Logic have played a role in the change. Anyway, even if a decisive landmark, Port-Royal Logic is still far from modern probability calculus and cannot be considered as the founding text. As James Franklin remarks, there is no serious calculation in this text as numbers are fictitious and decorative [13, p. 362]. Nevertheless, applying mathematical calculation to probability, and combining considerations on occurrence and on expectations, appear as two decisive steps towards probability calculus.

Finally, after fifty years of a difficult conceptualisation work, Bernoulli's *Ars Conjectandi*, written in 1689, published in 1713, is a much better candidate as the starting point of modern probability calculus. Bernoulli presents his text or *Ars conjectandi* as the successor of the Port-Royal Logic, which was named *Ars Cogitandi* by Arnauld and Nicole. Bernoulli's text starts where the previous Logic ended. We may discuss the effectiveness of this claimed heritage. However, this brings a new reason to look closely at the Port-Royal Logic as an important landmark in the emergence of probability calculus, and an invitation to revisit philosophical probability traditions and their impact on the construction of modern probability concepts.

The Port-Royal Logic mixes calculus of chance and philosophical probability; this mixed origin enlightens discussions on the nature of probability. Through those two traditions, two different issues and two different ways of dealing with uncertainty are addressed and linked together:

A moral issue: "What is the best decision under uncertainty? What should I do?" Where "best" is interpreted as "most rational" or "fair". To this line of thinking belongs the calculus of chances and its applications such as the division problem or the calculation of a fair price for risk-taking.

An epistemic issue: "How could I estimate if an uncertain proposition is true or not? What should I think? Is it probable or not?" Assuming that in itself it is true or false, at least in the XVIIth century. To this tradition belong Aristotle's Topics or the doctrine of probability of the XVIth to XVIIth centuries.

---

[5]First edition, 8 years after 1654.

In this paper we propose to explore the following questions: what are the epistemic issues around probability? How and when have epistemic and moral issues been connected?

2 About epistemic probability

If we tend to associate probability and random, or hazard, that was not the case in the early works about probability. Aristotle's concept of probability provides a good example of the explicit disjunction of the two ideas.

Indeed, Aristotle's probability has nothing to do with the idea of random or hazard. Those two notions correspond to τυχὴ (chance) or αὐτόματον (spontaneity) (*Phys.*, II.4–II.6). According to Aristotle, something takes place by chance if a person sets out with the intent of having one thing take place, with the advent of another unintended thing taking place. Spontaneity is what takes place by itself, without any rational preference and applies to animals and other beings. For instance, a horse walks spontaneously away from danger and saves its life, or a stool falls in such a way that we can sit on it. In both cases, the final cause and the efficient cause are disconnected, so there is no possible rational thinking about τυχὴ or αὐτόματον. For that reason, they are excluded from the realm of science and morality.

Probability is defined in the *Topica*. The word "probability" translates the Greek word ἔνδοξον, which is an opinion that is generally accepted. *Top.* I, part 1 defines probabilistic reasoning:

> Now reasoning is an argument in which, certain things being laid down something other than these necessarily comes about through them.
>
> 1. It is a 'demonstration', when the premises from which the reasoning starts are true and primary, or are such that our knowledge of them has originally come through premises which are primary and true;
> 2. reasoning, on the other hand, is 'dialectical', if it reasons from opinions that are generally accepted.

In that context, probabilistic reasoning is more or less synonym of dialectical reasoning and is not considered as a lower or less certain form of reasoning, as Aristotle confirms in Topics, book 1, part 2: we must say for how many and for what purposes the treatise is useful. They are three: intellectual training, casual encounters, and the philosophical sciences. There is still no connection of probability with scepticism or uncertainty, even if this connection was made during the Renaissance, when the Topics were rediscovered [11, p. xii].

When Aristotle's Topics are rediscovered during the Renaissance, commentators made a connection between δόξα and ἔνδοξα.[6] It led to the idea of a hierarchy of the knowledge. Science (ἐπιστήμη), would be the highest form of knowledge, what is true and necessary; and probable opinions, linked to the idea of δόξα, because of the name ἔνδοξα, would be what is doubtful and uncertain, a lower form of knowledge. However, there was no connection between δόξα and ἔνδοξα in Aristotle's works. This interpretation has been the origin of a new interpretation of the Topics during the Middle Age by Boëthius and Albert the Great.

Such an interpretation was also facilitated by the Latin translation of the Topics by Cicero as Marta Spranzi-Zuber has demonstrated [33]. Cicero translated ἔνδοξα to Latin by *probabilis* or *verisimilis*. Doing so, he translated three distinct concepts, applied to different things, by the same word:

- Ἔνδοξα are opinions which are generally accepted, beliefs shared by experts;

- Εἰκότα are facts happening most of the time, linked to the idea of regularity among natural phenomena;

- Πίστις is the quality attributed to an argument when reasoning, leading with more or less strength to a conclusion, which persuades the auditor, it represents the strength of a logical argument.

This tricky translation of ἔνδοξα transformed the idea of probability into a very polysemic notion, and made it a privileged concept for shifting from a meaning to another.

However, probability during the Middle Ages and the Renaissance is never connected to scepticism [14, pp. 120–125]. Even if we consider the various meanings of scepticism during the Renaissance, from Academic scepticism to pyrrhonism [21], in all cases, truth remains an objective reality, even if we cannot have access to it. The word "probable" suggest some connections between uncertainty and probabilistic ideas. In fact, uncertainty or probability may come either from the nature of the object we try to know, contingent and ephemeral, or from the quality of our agreement, or eventually from the persuading strength of the argument. In some cases, probability appears as a positive value in the Renaissance dialectic. In the context of a debate on a controversial issue, the doubt does not mean that truth would be inaccessible or contingent, but that we must continue to

---

[6]As defined in the *An. Post.*

search for it, and that the result of this research may not be definitive.[7] Probabilistic opinions do not imply any scepticism, even moderate, but expresses the variability and the contingent nature of our knowledge.[8]

Another decisive moment in the history of probability is the use of the concept by the casuists. Jesuit theologians (or casuists) proposed a new way of interpreting probability in a moral context during the Renaissance.

At issue here is the relation between knowledge and action: I need to know if an action is licit or illicit in order to act accordingly, or, if I am the confessor, in order to estimate to what extent an action is sinful or not. This kind of issue was very popular since the XIIIth century, when the *Summa confessorum* was flourishing. Significant examples of this literature are: *Summa de Paenitentia* by Raymond de Peñafort (ca. 1250) which will become the main source for *ulterior Summae* until ca. 1400, the *Summa juris canonici sive de casibus conscientiae* by Monaldus (1280), or the *Summa confessorum* by the Dominican friar John of Freiburg, used by other major Dominican theologians as Albert the Great or Aquinas. In this literature, in case of doubt, it is recommended to follow the most secure opinion that is *tutior*, in Latin. For this reason, this traditional way of dealing with uncertainty was called tutiorism. In most cases, if we are not sure if an action is licit or not, it is recommended to act in the most secure way, i.e., as if it was not licit, and so to restrain from doing it, in order to avoid sin, as sin may impair chances to be saved.

Jesuit theologians challenged Tutiorism during the Renaissance. Instead of recommending following what is the most secure, they recommended to follow what is the most probable or probabilior, even if there is some possibility that it could be a sin. At first, they recommend to follow only what is "probabilior", very probable or almost certain as Bartolomeo de Medina. Later, other theologians recommend following an opinion even if it is only "probabilist" or probable and even a very controversial opinion, as Suarez. Therefore, they were called "probabilists".[9]

---

[7] *Le caractère contingent de la vérité d'une proposition n'affecte pas la valeur de vérité de la proposition elle-mme, mais concerne notre relation à cette valeur de vérité. L'assentiment accompagné de doute dépend donc de la conscience que notre connaissance n'est que conjecturale et révisable, même si le but ultime de toute connaissance reste la correspondance entre nos croyances et la vérité.* [33, p. 310].

[8] This issue has been discussed. Some contemporary authors connect the revival of scepticism to the emergence of probability, as Lorraine Daston in Probability in the Enlightenment, some XVIIth century authors as Arnauld accuse probabilism to open the road to scepticism. The connection between XVIth century scepticism and XVIIth probabilism is true, but implies an evolution of concepts. On the reverse, during the XVIth century itself, probabilism and scepticism are not connected because they address different issues.

[9] On probability in theology, cf. [12, 30].

In the moral theology of the XVIth century, it is possible to doubt and to hesitate between two probable opinions, but as soon as one of the two opinions is acknowledged as the more probable by the mind, the mind can no longer consider both opinions as probable, and has to choose the more probable. In the new Jesuit theology, two opinions may remain probable, which means that the agreement of the mind is not important anymore. Arnauld and Pascal will criticize this attitude. For them, far from being a sign of subjectivism, the agreement of the mind is an effect of the realism of truth. Without any necessary acknowledgement of the mind, probable opinions become autonomous and are disconnected from the truth. In that case, they become independent of the truth and therefore contingent and exposed to subjectivism.

In order to solve the impasse of simultaneous probable opinions, Gabriel Vasquez introduced a distinction between "intrinsic" and "extrinsic" probability in his *Commentaria ac disputations* (1597) on Aquinas's Summa Theologica. This division reminds more or less the *"circonstances extérieures"* and *"circonstances intérieures"* in the Port-Royal Logic. For Vasquez, "intrinsic probabilities" are the opinions we accept, according to our own convictions, intrinsically justified, even if the opposite opinion may exist and be defended by other doctors. The opposite opinion is considered therefore as "extrinsically probable", and in general, extrinsic probabilities are close to the opinions generally shared by experts and doctors (which was the original meaning of ἔνδοξα).

This theory allows simultaneous probabilities, no matter whether one of them does not produce an acknowledgement of the mind. Even if Vasquez preserves the idea that action must be conform to objective truth and recommends to follow the most secure opinion as did the tutiorists, the idea of simultaneous probability leads to a focus on extrinsic probabilities defended by other doctors in the casuistic literature, and this opens the floor to new probabilistic books quoting the names of various doctors. We know it through the sharp critics addressed to them by Pascal in the *Provinciales*.

Francisco Suarez (around 1600) introduces a new criterion in the evaluation of probability: practical certainty. According to him,[10] rational persuasion occurs either *aliqua auctoritate* (by extrinsic probability, the opinions of doctors), either by *ratione* (reason is used to determine intrinsic probability), or *per sufficientia principia practica* (by sufficient practical principles). This third criterion may be considered as a first expression of "what we can expect" from a given situation, a first attempt to calculate rationally what we can expect here and now. The reference to the temporality of the action, instead of the usual reference to universal truth, introduces a new

---

[10]In *De Bonitate et Malitia Humanorum Actuum*, XII.iv.iv; quoted by Deman [12].

way of thinking about decision making under uncertainty, not so far from probability calculus.

Suarez also proposes a new way of thinking about doubt. Before him, doubts concerned our ability to discover the law and the objective truth, universality and reality of the law were never questioned. Suarez admits that doubt may also concern the existence of the law itself, in that case, if there is no law, there is no moral obligation, and consequently no obligation to choose tutiorism. If there is no law, there is no doubt about our ability to discover it to conform our actions to it, and therefore we are facing only probable opinions. To choose among them, we will use the three criteria mentioned above. Instead of following the most secure opinion, we will follow the most probable.

This is a completely innovative way of connecting action and knowledge. We are very close to the situation exposed in the wager: doubt is equivalent to ignorance. Making a decision under uncertainty means choosing the best ratio between what we can expect from a given situation and what we can fear, or what seems the more licit according to moral law. In other words, the decision is made according to what appears the most probable according to your judgement. This structure is rather similar to the structure of action under uncertainty in the Port-Royal Logic.

Still, without any scheme to calculate probability and to compare intrinsic and extrinsic probabilities, this structure was exposed to a subjective evaluation of probability according to personal wishes, and therefore, to relativism. Another difference with the conception of action proposed in the Port-Royal Logic is the application of probabilities to religious decisions, in relation to salvation, when Arnauld and Nicole apply probability only to civil matters, without any religious involvement.

Published after the publication of the Port-Royal Logic, it is interesting to mention here the work of Gonzalez de Santalla, *Fundamentum theologiae moralis id est tractatus theologicus de recto usu opinionum probabilium*, published in 1694. As in earlier theological books, probability is defined once more in relation to truth: *vir prudens judicet rem esse veram* and to personal agreement: *ob rationem vel rationes talem prae se ferentes apparentiam veritatis* [15, p. 11]. Because the opinion appears as a true opinion to the mind, it produces the rational agreement of the mind to this opinion.

In the second part of the treaty, Gonzalez de Santalla describes a renovated version of the doctrine of probability called "probabiliorism". He recommends to make decisions under uncertainty considering what is *clare et sensibiliter verissimilior* [15, p. 125]. This new solution reconciles the quality of the agreement of the mind, depending on the apparent truth of the opinion, with the freedom to interpret rationally a given situation. The same kind of middle way between dogmatism and scepticism is shared with other supporters of a probabilistic logic during the same period.

The influence of theological probability would require further investigations in order to clarify the role it played in the construction of the concept of probability during the XVIIth century. If the former moral theology is challenged when tutiorism and personal agreement to an objective moral law are contested, both are finally restored in a new conceptual construction of action. In the meantime, the discussion about probability had opened a space for new connections between personal rational choice under uncertainty and knowledge of the truth. In that respect, philosophical (and theological) probability certainly contributed to the emergence of modern probability calculus along other traditions as calculus of chances.

We will examine now how the Port-Royal Logic mixed the calculus of chances and the ideas from the doctrine of probability, despite the fact Arnauld is one of the most passionate polemicist against it.

3   The connection of epistemic and moral issues about probability in the Port-Royal Logic

The fourth part of the Port-Royal Logic[11] (1662-1683) by Antoine Arnauld and Pierre Nicole represents probably one of the first attempts to connect calculus of chance and probability. For the first time, the word "probability" is associated to calculus on random games. At the same time, the part four uses the casuist vocabulary to think epistemic probability with notions as human faith, testimonies, etc. The connection to the calculus of chance is also explicit with a mention of Pascal works in the last chapter and lottery taken as an example in chapter xvi.

The five last chapters of the Port-Royal Logic are dedicated to this issue. Chapter 12 define types of faith: what we know by faith, whether human or divine. Chapters 13, 14, 15 are dedicated to past events known by human faith. Chapter 16 concerns the judgments we ought to make concerning future accidents. These rules, which are helpful for judging about past events, can easily be applied to future events. The parallel between past and future events allows the connection between game calculus and probable opinions.

In chapter 12, two types of faith are defined:

> For there are two general paths that lead us to believe that something is true. The first is knowledge we have of it ourselves, from having recognized and examined the truth either by the senses or by reason. This can generally be called reason, because the senses themselves depend on a judgment by reason, or science, taking this name here more generally than it is taken in the Schools, to mean all knowledge

---

[11]The first edition was published in 1662, then there were four augmented editions till the fifth in 1683. We used the French edition [2] and quote the English translation [3].

> of an object derived from the object itself. The other path is the authority of persons worthy of credence who assure us that a certain thing exists, although by ourselves we know nothing about it. This is called faith or belief [...]. But since this authority can have two sources, God or people, there are also two kinds of faith, divine and human. [3, Part iv, Chapter 12, p. 260]

This typology determines a typology of knowledge and subjective modalities. Following Claude Imbert, four different models may be identified, where belief (or credence) is determined by a measure, which leads to decision-making or controls it [17]. This typology defines the outline of the next chapters. In chapter 12, absolute certainty and absolute uncertainty are defined. Absolute certainty is divine certainty as God makes us believe infallibly, absolute uncertainty is represented by human authority, always deceiving and deceived. However, those two extremes are not the focus of the next chapters where the scope is what is not completely certain neither uncertain. Chapters 13 to 15 deal with past and future events and defines criteria to estimate to what extent we should believe in them or not. Internal and external circumstances are the criteria to estimate past events, fears, desires and probability of occurrence must be combined to estimate future events. Chapter 16 ends with a critical comment of the Pascal's wager where, according to Arnauld, no relation can be thought between desire and expectations because of the divine nature of the issue. Obviously, estimation of degrees of belief relates only to human faith and human affairs. Divine faith is excluded of a possible calculus.

In chapter 13, entitled "Some rules for the directing reason well in beliefs about events that depend on human faith", a rule to estimate if I should believe a thing or not is explained:

> In order to decide the truth about an event and to determine whether or not to believe in it, we must not consider it nakedly and in itself, as we would in a proposition in geometry. But we must pay attention to all the accompanying circumstances, internal as well as external. I call those circumstances internal that belong to the fact itself, and those external that concern the persons whose testimony leads us to believe it. [3, Part iv, Chapter 12, p. 264]

The mention of "external" and "internal" circumstances reminds us of "extrinsic" and "intrinsic" probabilities in Vasquez's works, but in the Port-Royal Logic, it does not apply to the judgement made upon a thing, but to the thing itself. The sentences that follow immediately this text contain also a clear reference to the issues related to the agreement of the mind and its natural propensity to believe what is objectively true, a topic we have discovered in the casuistic literature:

> Given this attention, if all the circumstances are such that it never
> or only rarely happens that similar circumstances are consistent with
> the falsity of the belief, the mind is naturally led to think that it
> is true. Moreover, it is right to do so, above all in the conduct of
> life, which does not require greater certainty than moral certainty,
> and which even ought to be satisfied in many cases with the greatest
> probability. [3, Part iv, Chapter 12, p. 264]

If Arnauld criticizes the interpretation of probability by some casuists, he
uses the same structure to describe action as a choice determined by a belief
or a probable opinion, but this decision-making structure is shifted from
theology to everyday life. It is also noteworthy to remark that the word
"probability" is employed to measure the moral certainty of those actions.

At the same time, Arnauld thinks action as a choice, as a decision according to the most probable. In doing so, he relies on the same structure
as the calculation of decision making under uncertainty, for which random
games are a model. Therefore, the introduction of "moral certainty" and
"probability" changes the framework of decision making. If we compare this
model to Pascal's model, in Pascal's problems, there is no way to know if an
event is true or not (the event is a contingent future event). Because of this
fundamental ignorance, Pascal proposes to calculate the decision by considering chances of loss and chances of gain, and to make a rational decision,
but he never proposed to examine the probability of the possible events,
which is, by definition, impossible to know in his model.

By joining probability to game calculus, Arnauld proposes here an important shift, from rational decision making towards estimation of an epistemic value, which is a new way of conducting the actions of our life. The
scheme that allows such a shift is the equivalence between past and future
events; even though they should be considered of different ontological nature as past events are 8 compared here to contingent future events. In
fact, Arnauld does not consider the events themselves, but our judgements
about the events, consequently, he can compare judgements and apply to
them the same rules, as explained in chapter 16, entitled "Of the judgment
which ought to make concerning future accidents":

> The rules, which are helpful for judging about past events, can easily
> be applied to future events. For, as we ought to believe it probable
> that an event has happened whenever certain circumstances we know
> about are ordinarily connected with the event, we also ought to be-
> lieve that it is likely to happen whenever present circumstances are
> such that they are usually followed by such an effect. [3, p. 273].

Arnauld operates here a kind of double shift: On one hand, what is probable is not the event itself anymore, but the probability to occur, or more

precisely, our judgement about the probability the event occurs, which is a change compared to the Jesuit doctrine of probability. We also remark that the "opinion" has become "judgement" about an "event". He substitutes to the epistemic question "did the event happen?" a moral question: "What should I do?" in a connection that seems at first rather similar to Pascal's wager, however quite different. In Pascal's texts, the probability of future events is unknown by definition and not even mentioned[12], which is not the case here.

On the other hand, decision under uncertainty is here strictly limited to human affairs when what was at stake with Pascal or the casuists was eternal life. The application domains are in fact very different as well as the epistemic constructions. The decision issue "what should I do?" and the epistemic issue "what can I know?" were completely disjointed in Pascal's works, it is connected here in a new way, in order to avoid the Jesuit relativism, according to Arnauld.

After the framework has been exposed, Arnauld explains how to compose the consequences and the probability of occurring:

> [...] many people happen to fall into an illusion that is all the more deceptive as it appears more reasonable to them. This is that they consider only the greatness and importance of the benefit they desire or the disadvantage they fear, without considering in any way the likelihood or probability that this benefit or disadvantage will or will not come about. So, whenever they are apprehensive about some great harm, such as loss of life or all their wealth, they think it prudent not to neglect any precaution to safeguard themselves against it. And if it is some great good, such as gaining a hundred thousand crowns, they think they are acting wisely by trying to obtain it if the risks are slight, however unlikely they are to success. [3, p. 273]

Rational decision, according to Arnauld is not taken according to the most advantageous proposition nor the most likely to occur, but by combining both in a new calculus. Without this combination, people would fall into excess of fear, as the princess who fears climbing to the first floor because the floor could collapse under her feet, or into excess of optimism as the gamblers who bet at the lottery. The example of the lottery gives the opportunity to apply calculus of chance to determine if the game is fair. A fair game will be here the norm of the rational decision we should make. Where the tutiorist moral would recommend to follow the most secure, in human affairs, Arnauld proposes to calculate degrees of probability by combining the utility and the probability in a calculus inspired by calculus of chances:

---

[12]See the division problem in the *Traité du triangle arithmétique* or the wager in the Pensées [6, no. 233].

The flaw in this reasoning is that in order to decide what we ought
to do to obtain some good or to avoid some harm, it is necessary to
consider not only the good or harm in itself, but also the probability
that it will or will not occur, and to view geometrically the proportion
all these things have when taken together. This can be clarified by
the following example. There are games in which, if ten persons each
put a crown, only one wins the whole pot and all the others lose.
Thus each person risks losing only a crown, and may win nine. If we
consider only the gain and loss in themselves, it would appear that
each person has the advantage. But we must consider in addition
that if each could win nine crowns and risks losing only one, it is
also nine times more probable for each person to lose one crown and
not win the nine. Hence each has nine crowns to hope for himself,
one crown to lose, nine degrees of probability of losing a crown, and
only one of winning the nine crowns. This puts the matter at perfect
equality (in French. *Ce qui met la chose dans une parfaite égalité*).
[3, pp. 273–274]

Instead of proposing directly a calculus of the expectation as Huygens
or Pascal, Arnauld creates a new quantitative variable, which is probability.
He distinguishes potential gain or loss, nine crowns to hope for himself, one
crown to lose, and the degrees of probability to gain or win, nine degrees
of probability of losing a crown, and only one of winning the nine crowns.
Those degrees of probability are not what we call probability, a number
between 1 and 0, defined as the number of successful events on the total
number of events. Arnauld only defines here a double relation between gains
and probabilities. If the products of degrees of probability by the gains are
equal on both sides, then the game is fair, if not, it is unfair.

What is important in decision making according to Arnauld is the proportion between danger and advantages:

> ...since it is neither by the danger nor the advantages, but by the
> proportion between them that we should judge [the accidents]. [3, p.
> 275]

In an example about the apprehension some people have towards thunder,
he makes another attempt to quantify the danger, estimating that one on
two million persons may be killed by thunder. Such a consideration opens
the floor to a sort statistical thinking about accidents, as we find also in
political arithmetic. Acting rationally is not anymore measuring separate
quantities, gain or loss and respective probability to occur, but calculating
the proportion between the two. This calculus makes us escape to the
irrationality of feelings and impressions and should be applied to all domains
concerning human affairs: "The main use we ought to derive from them is
to make us more reasonable in our hopes and fears." [3, p. 274] It defines a
new morality in human affairs.

This new method applies only to human affairs, not to divine matters. It works only with finite things:

> Only infinite things such as eternity and salvation cannot be equalled by any temporal benefit. Thus we ought never balance them off against anything any worldly. This is why the slightest bit of help for acquiring of salvation is of worth more than all the goods of the world taken together. And the least peril of being lost is more important than all temporal harms considered merely as harms. [3, p. 275]

When Pascal's wager is based on a calculus on the infinite, Arnauld refuses to apply calculus of probability to the infinite, even to the mathematical infinite. Even if he reproduces partly Pascal's argumentation of the infinitely small (part iv, chapter 1), the infinite is excluded from the field of logic. If calculus apply in ordinary life, they are useless concerning eternal life where the traditional tutiorist moral remains the standard. In fact, Arnauld's argument is different of Pascal's regarding the ability to choose eternal life. For Pascal, choosing eternal life or not may be the result of a choice by a rational subject, who can use a mathematical calculus to determine himself. For Arnauld, in such matters, there is no possible rational choice when divine faith is at stake.

4   Conclusion

A history of probability calculus should not neglect the philosophical concept of probability, the doctrine of probability initiated by Jesuit theologians and the Port-Royal Logic probability as they played an important role in the emergence of probability. The Port-Royal Logic should be considered as a landmark, as it is the place where epistemic probability is connected to the calculus of chance in an original construction of the rational action that remains more or less until today.

In the fourth part of the Port-Royal Logic, Arnauld sets the foundations of a new logic for everyday life, which will be developed later by Jakob Bernoulli in the Ars conjectandi, presented as the continuation of Arnauld and Nicole's *Ars cogitandi*.[13] In this text, Arnauld designs a new conceptual framework characterized by the following shifts.

First, calculus of chance, which used to be oriented towards decision making under uncertainty, where uncertainty was impossible to circumvent by definition, becomes epistemic-oriented through the equivalence of past and future events. Second, irrational fears and hopes, are replaced by a rational proportion between probabilities and utilities, in a rational attempt

---

[13]This statement could be discussed, but is explicitly expressed by Bernoulli himself in the introduction to the *Ars Conjectandi* [19, 34]. On the role of Jakob Bernoulli for the development of probability theory, cf. [24].

to avoid dogmatism, unable to deal with contingent facts, and scepticism. Third, from a logic based on categories he shifted to a logic based on representations. In this space, it became possible to build a methodology to measure "degrees of probability", from absolute certainty to absolute doubt, without questioning the objective reality of truth, and therefore to avoid scepticism. Fourth, probability which used to be an epistemic qualitative notion, and later a moral theological doctrine, was transformed into a calculus for everyday life, with possible quantifications, even if the examples are still approximate. Probability adopted concepts from calculus of chance and from qualitative becomes quantitative.

Bibliography

[1] Antoino Arnauld and Pierre Nicole. *Logic, or the Art of Thinking.* James Gordon, Edinburgh, 1861. Translated from the French by Thomas Spencer Baynes.

[2] Antoine Arnauld and Pierre Nicole. *La Logique ou l'art de penser (1662).* Vrin, Paris, 1981. Edited by Pierre Clair and François Girbal.

[3] Antoine Arnauld and Pierre Nicole. *Logic or the Art of Thinking: the Port-Royal Logic.* Cambridge University Press, Cambridge, 1996.

[4] Evelyne Barbin and Pierre Lamarche, editors. *Histoire de probabilités et de statistiques.* Ellipses, Paris, 2004.

[5] Bernard Bru and René Thom, editors. *Pierre-Simon Laplace. Essai philosophique sur les probabilités (1814).* Bourgois, Paris, 1986.

[6] Leon Brunschwicg, editor. *Blaise Pascal. Pensées.* Librairie Generale Francaise, Paris, 1972.

[7] Edmund Byrne. *Probability and Opinion: A Study in the Medieval Pre-suppositions of Post-Medieval Theories of Probability.* Martinus Nijhoff, Den Haag, 1968.

[8] Ernest Coumet. Le problème des partis avant pascal. *Archives internationales d'histoire des sciences*, 18(72–73):245–272, 1965.

[9] Ernest Coumet. La théorie du hasard est-elle née par hasard? *Annales: Economie, Sociétés, Civilisations*, 3:575–598, 1970.

[10] Ernest Coumet. *Sur le "calcul ès jeux de hasard" de Huygens: dialogues avec les mathématiciens français*, pages 123–137. Vrin, Paris, 1982. Foreword by René Taton.

[11] Lorraine Daston. *Classical Probability in the Enlightenment*. Princeton University Press, Princeton, NJ, 1988.

[12] Thomas Deman. Probabilisme. In Alfred Vacant, Eugène Mangenot, and Emile Amann, editors, *Dictionnaire de théologie catholique*, pages 417–622. Librairie Letouzey et Ané, 1936. Tome XIII.

[13] James Franklin. *The Science of Conjecture. Evidence and Probability before Pascal*. John Hopkins University Press, Baltimore, 2001.

[14] Anne-Sophie Godfroy-Genin. *De la doctrine de la probabilité à la théorie des probabilités. Pascal, la Logique de Port-Royal, Jacques Bernoulli*. PhD thesis, Université Paris 4, Sorbonne, 2004.

[15] Tirso González de Santalla. *Fundamentum theologiae moralis id est tractatus theologicus de recto usu opinionum probabilium*. Sumptibus Gasparis de Stortis, Venice, 1694.

[16] Ian Hacking. *The Emergence of Probability*. Cambridge University Press, Cambridge, 1975.

[17] Claude Imbert. Port-Royal et les modalités subjectives. *Les temps de la réflexion*, 3, 1982.

[18] Leonid Maistrov. *Probability Theory: A Historical Sketch*. Academic Press, New York, 1974. Translated from the Russian and edited by Samuel Kotz.

[19] Norbert Meusnier, editor. *Jacques Bernoulli, Ars Conjectandi (1713)*. Institut de Recherche sur l'Enseignement des Mathématiques, Rouen, 1987. Text in Latin and translation.

[20] Siméon Denis Poisson. *Recherches sur la probabilité des jugements en matière criminelle et matière civile*. Bachelier, Paris, 1837.

[21] Richard Popkin. *The History of Scepticism from Erasmus to Spinoza*. University of California Press, Berkeley, 1979.

[22] Ivo Schneider. The Contributions of the Sceptic Philosophers Arcesilas and Carneades to the Development of an Inductive Logic Compared with the Jaina-Logic. *Indian Journal for the History of Science*, 12:173–180, 1977.

[23] Ivo Schneider. Why Do We Find the Origin of a Calculus of Probabilities in the Seventeenth Century? In Jaakko Hintikka, David Gruender, and Evandro Agazzi, editors, *Probabilistic Thinking, Thermodynamics*

and the Interaction of the History and Philosophy of Science. Volume II. Proceedings of the 1978 Pisa, Italy, September 4–8, 1978 Conference on the History and Philosophy of Science, volume 146 of Synthese Library, pages 3–24, Dordrecht and Boston, 1980. Reidel.

[24] Ivo Schneider. The Role of Leibniz and Jakob Bernoulli for the Development of Probability Theory. LLULL Boletín de la Sociedad Española de Historia de las Ciencas, 7:69–90, 1984.

[25] Ivo Schneider. Die Entwicklung der Wahrscheinlichkeitstheorie von den Anfängen bis 1933. Einführung und Texte. Wissenschaftliche Buchgesellschaft, Darmstadt, 1988.

[26] Ivo Schneider. The Market Place and Games of Chance in the 15th and 16th Centuries. In Cynthia Hay, editor, Mathematics from Manuscript to Print 1300-1600, pages 220–235, Oxford, 1988. Oxford University Press.

[27] Ivo Schneider. Christiaan Huygens' Non-Probabilistic Approach to a Calculus of Games of Chance. De Zeventiende Eeuw, 12(1):171–185, 1996.

[28] Ivo Schneider. Abraham De Moivre, Doctrine of Chances (1718, 1738, 1756). In Ivor Grattan-Guinness, editor, Landmark Writings in Western Mathematics 1640-1940, pages 105–120, Amsterdam, 2005. Elsevier.

[29] Ivo Schneider. Jacques Bernoulli, Ars Conjectandi (1713). In Ivor Grattan-Guinness, editor, Landmark Writings in Western Mathematics 1640-1940, pages 88–104, Amsterdam, 2005. Elsevier.

[30] Marta Gracia Secades, Javier Martin Pliego, and Jesús Santos del Cerro. L'influence du probabilisme espagnol sur la conceptualisation de la probabilité, 2002. Séminaire d'histoire des probabilités et de la statistique, Centre Koyré, EHESS, Paris.

[31] Philippe Sellier, editor. Blaise Pascal. Pensées. Garnier, Paris, 1991.

[32] Oskar Sheynin. On the Prehistory of the Theory of Probability. Archives for History of Exact Sciences, 12(2):97–141, 1974.

[33] Marta Spranzi-Zuber. Rhétorique, dialectique et probabilité au xvie siècle. Revue de Synthèse. 4e série, 122(2–4):297–317, 2001.

[34] Bing Sung, editor. Jacob Bernoulli, Ars Conjectandi. Translation of Part IV. Includes the Correspondence between Leibniz and Bernoulli.

1966. Harvard University, Department of Statistics. Technical Report No. 2.

[35] Isaac Todhunter. *A History of the Mathematical Theory of Probability from the Time of Pascal to that of Laplace (1865)*. Chelsea Pub. Co., New York, 1949.

Benedikt **Löwe**, Eric **Pacuit**, Jan-Willem **Romeijn** (*eds.*)
Foundations of the Formal Sciences VI
Reasoning about Probabilities and Probabilistic Reasoning

Measuring the Uncertain: A concept of objective single case probabilities

MARTIN NEUMANN*

Institute for Philosophy, Bayreuth University, Universitätsstraße 30, 95447 Bayreuth, Germany
E-mail: `martin.neumann@uni-bayreuth.de`

1 Introduction: a basic primitive concept for probabilities

This essay re-investigates an interpretation of the probability calculus developed in 1886 by the German physiologist Johannes von Kries: the theory of so called *Spielräume* (i.e., range or scope). For many decades his 'Principien der Wahrscheinlichkeitsrechnung' remained the most influential book on probability theory in the German language and has been regarded as the most elaborate interpretation of the probability calculus. This suggests that interesting ideas can be found in this account, which—to a certain degree—might even be able to shed light on the historical contingency of the present.

In the 19th century, applications of probability theory gained more and more attention in scientific practice, ranging from error theory [60] to social statistics [25, 50, 23, 18]. Due to the kinetic theory of gases it even invaded physics [55, 58, 73]. The 19th century was faced with a probabilistic revolution [33]. In particular, at the beginning of the 19th century, statistical data were collected all over Europe [23]. This provided new means to determine cases of equal possibility by counting relative frequencies. Around 1840, several authors, including Mill [39], Ellis [11] and Fries [15] emphasised this idea. However, 19th century German discourse on probability was highly sceptical about statistical laws [53]. Investigations into why Statis-

---

*I would like to thank participants of the FotFS VI conference and an anonymous referee for fruitful and encouraging discussions, comments, and recommendations. I would like to thank Teresa Gehrs for improving my English.

Received by the editors: 27 July 2007; 19 December 2007.
Accepted for publication: 31 January 2008.

tics was rejected in Germany can be found in [22, 23, 49, 41]. However, this probabilistic revolution of scientific argumentation has led to intensive discussion of the foundations of probability theory itself. The more the probability calculus was applied, the more it became unclear what it was about. This is the context in which Johannes von Kries *Spielraum* theory of probability became popular. In modern terms, the aim of this theory can be characterised as giving a *basic primitive concept* for the foundation of probability not in terms of probability itself. For example, the notions of relative frequencies or of the relation of preference structures are basic primitive concepts in the frequentist, respectively the subjectivist, conception of probability. *Spielräume* are conceived of as such a basic primitive concept. Thus, this account has to develop a physical categorial basis F [37] that enables probabilistic reasoning for single cases. By developing the notion of *Spielraum* von Kries found such a basis.

Von Kries illustrates his main idea using the example of a meteorite hitting the earth [67, p. 24]. What is the probability that the meteorite will hit a certain district? If it is only known that the meteorite will hit the earth, but not exactly where, the probability that the meteorite will hit Europe or Africa is nevertheless quite different. Africa is much bigger than Europe. To be unequivocal, the division of the surface of the earth has to be worked out according to sound procedures. Von Kries developed measurement protocols to guarantee that probability is proportional to the size of the area[1]. He concluded that the application of the probability calculus is possible if "our assumptions are concerned with an item, for which, according to our knowledge, it seems possible that a measurable and decomposable range of behaviour exists" [67, p. 24]. The further elaboration of this idea led von Kries to identify a physical categorial basis for the probability calculus.

This example already clearifies that von Kries was concerned with the probability of a single event. His *Spielraum* theory claimed that so-called *Spielräume* are objectively determined single cases. At first glance, this looks like modern propensity theories. However, it turns out that fundamental differences remain with regard to positions typically related to modern propensities. Most importantly, propensities are typically measured by relative frequencies and are closely related to an indeterministic world view. The modern Propensity Interpretation was introduced by Karl Popper in his influential articles of 1957 and 1959. Subsequently a variety of different

---

[1]This example can be subsumed under so-called geometrical probabilities. Geometrical probabilities were popular in the late-19th Century. The best known example is the Bertrand Paradox, which examines the probability that a line drawn randomly through a circle will be greater than a triangle in the circle. This example is similar to the Bertrand Paradox. Kamlah introduces the notion of a von Kries-Bertrand Paradox, since von Kries' work was prior to Bertrand's [28, 29].

concepts of propensities have been introduced. For instance, long-term and true single case propensities can be distinguished [19]. Thus, the *Spielraum* concept cannot be contrasted to a single canonical concept of Propensities, however crucial differences remain. These can be summarised in the following two points:

Propensities and relative frequencies

First, the observation of physical propensities needs to be considered. According to Popper's classical concept of propensities, they represent theoretical entities that should be measured by relative frequencies. In discussing the case of throwing a 6 in the game of die, he writes:

> We may agree to take as our measure of that propensity the (virtual) *relative frequency* with which the side turns up in a (virtual, and virtually infinite) sequence of repetitions of the experiment. [47, p. 32].

Propensities are features of an experimental arrangement that, in the long, run creates a limit of relative frequencies. They are a property of the generating conditions of a given sequence. There is, thus, a strong commitment to frequentism in this account: Although probability is not defined by relative frequencies, actual propensity values are measured by them [46]. This account has been contested by Fetzer [13] and Miller [40]. Later on, Popper himself changed his mind [48]. In particular, these authors object that propensities are not a measure of any frequency. Yet the question of how to determine concrete propensity values remains problematic. For example, in his later work on 'a world of propensities' Popper himself simply referred to "estimating them speculatively" [48, p. 17].

Propensities and indeterminism

Moreover, modern accounts of propensity theory commonly refer to an ontic indeterminism. Popper was convinced of the existence of physical propensities by investigating the interpretation of quantum theory [45]. Quantum mechanics is the paradigmatic example of indeterministic laws of nature. Therefore, according to Popper, propensities are "physical properties comparable to Newtonian forces" [46, p. 27]. He thus assumes that propensities are law-like dispositions or tendencies that can be found in nature. At any rate, as Giere puts it, "if we knew our world to be deterministic, discussions of single-case probabilities would be quite academic." ([17, p. 475]; see also [16]) Thus, propensities are considered as an indeterministic generalisation of the concept of causation or causal probabilities [54, 19].

In conclusion, modern propensity theories are closely related to the following two assumptions; namely:

1. Propensity values can be measured by recourse to *relative frequencies*.

2. Propensity values other than 0 or 1 can only arise in an *indeterministic* context.

In the following discussion, it will be demonstrated that both assumptions do not hold for the theory of *Spielräume*: Contrary to the first assumption *Spielräume* represent an attempt to develop a process of measurement for objective single case probabilities. Contrary to the second assumption, the *Spielraum* theory was developed to reconcile probabilistic argumentation with deterministic Newtonian mechanics. Hence, the historical example of the *Spielraum* concept of probability can serve as a proof of existence that these elements are *not necessary* elements of objective single case probabilities.

To unfold this argument, the current paper proceeds as follows: It consists of two main parts. The first addresses the question of how to measure single case probabilities. That is, it deals with the relation of *Spielräume* to relative frequencies. The second part investigates the problem concerning the relation between probabilism and determinism.

Both questions, however, call for an outline of the historical context. The first section is more technically oriented: the question of measurement has to be understood in terms of the contemporary framework; namely, the Laplacian probability theory. One of the thorniest technical problems for this theory has been the determination of equal possible cases. This problem will be investigated in its historical setting, before turning attention to the solution to this problem proposed by Johannes von Kries' *Spielraum* theory. This section will work through two stages: first, a game of chance will be introduced, and then an extensive measurement structure will be developed that will be illustrated by this game. The second part of the paper builds upon these results and investigates more philosophical questions connected with the meaning of probabilistic statements. These will be elaborated within the contemporary scientific context. The development of the kinetic theory of gases was the first entry of chance into the deterministic framework of classical mechanics. Therefore, a brief outline of the debates associated with the kinetic theory of gases will first be given, before von Kries' contribution to this development is investigated in more detail. It will be shown that the measurement structure outlined in the first part of this paper appeared to be fulfilled in this case. This framework allowed von Kries to reconcile probability with determinism.

2   First part: measuring objective single case probabilities

As a first step, the question of how objective single case probabilities can be measured according to the *Spielraum* theory will be analysed. Since this theory still stands in the tradition of Laplacian probability theory, answering the question entails re-examining the framework of Laplacian probability theory. On the one hand it has to be clarified how the problem of determining probability values was shaped by the Laplacian framework, and on the other, why and how von Kries sought objective probabilities within this framework.

2.1   Equipossibility in Laplacian probability theory

The *Spielraum* theory aimed to solve a fundamental technical problem of Laplacian Probability Theory: namely, the determination of cases of equal possibility. This problem was determined by the definition of probability given by Laplace. Within this framework, probability was defined as the quotient of the favourable and all possible cases:

$$\text{Probability} = \frac{\text{Number of favourable cases}}{\text{Number of possible cases}}$$

This definition can already be found in the work of Bernoulli [1]. Other predecessors of Laplace similarly relied on this definition [21]. The paradigm for probabilistic reasoning were games of chance such as throwing a die. If one throws an unbiased die once only, there are obviously six possible outcomes. However, the division of favourable by possible cases makes sense only if all cases are of equal possibility. Already in the case of repeated games of chance, the number of possible cases is no longer so obvious. The most prominent counterexample was the famous riddle of D'Alembert: Peter and Paul play with a coin. Peter wins if the coin falls heads on the first or the second toss. D'Alembert then asks: what is the probability that Peter wins the game? With H for head and T for tail there are four possible cases, namely: (H, H), (H, T), (T, H), (T, T). However, D'Alembert counted (H, H) and (H, T) as one case because if head falls first the game is terminated. Thus, he distinguished only three cases. D'Alembert's (incorrect) solution shall not be discussed here [60]; it need only be remarked that the problem of determining equally possible cases was one of the central objections to equipossibility theories of probability from the outset. Laplace himself remarks that this is "one of the most delicate points in the theory of chances" [10, p. 6].

Throughout the 19th century the subject formed one of the most urgent topics in debates associated with the foundations of probability theory, at least until equipossibility theories of probability were finally abandoned in

the beginning of the 20th century [29, 73]. Von Kries concluded that it was
"very often noticeable that a considerable difficulty is found in the determination of the equally possible" [67, p. 274]. The dominant theory within
German discourse was the so-called theory of 'logical ignorance'. According
to this theory, two cases are of equal possibility if nothing can be said in
favour of one or the other case. Most of the contemporary logical textbooks
in Germany insisted on this so-called principle of insufficient reason,[2] today
known as indifference principle in Bayesian accounts. For the purposes of
scientific application it was closely connected to a metaphysics of causes
[23, 29]. If, e.g., the probability of a male birth is 108 : 100 this theory
states that there are 108 causes in favour and 100 causes against a male
birth [61]. One therefore knows that there are 208 different causes and that
one is in a state of complete ignorance about the actually working causes.
Hence, this essentially represents a subjective interpretation of possibility.

2.2   Objective probabilities and logical ignorance

However, the rise of science demanded objectivity. In particular, the logical
foundation failed to convince von Kries since the determination of equally
possible cases has to be unequivocal. Such a determination cannot be se-
cured solely by the principle of insufficient reason. To demonstrate this, he
presented the following counterexample:

One can pose the following question: is there a terrestrial chemical element on a foreign star like Sirius [67, p. 10]. If there were no research
instruments to answer the question empirically, one would be in a state of
total ignorance. Therefore, according to the theory of logical ignorance, it
is of equal possibility that there is iron on Sirius and that there is no iron
on Sirius. Thus, the probability for iron on Sirius is $\frac{1}{2}$. The same argumentation holds for every chemical element. The multiplication theorem then
tells us that the probability that there is neither iron nor (say for example)
gold to be found on Sirius is $\frac{1}{2} \times \frac{1}{2} = \frac{1}{4}$. Iterated application of the multiplication theorem leads to a very high probability that there is some terrestrial
chemical element on Sirius. Yet, one could also pose directly the question
of whether there is some terrestrial chemical element on Sirius. Following
the theory of logical ignorance one has to say that there is nothing known
in favour of this or the opposite and therefore the probability has to be
determined as $\frac{1}{2}$. Hence, the determination of the probability value is not
unequivocal. Von Kries draws the conclusion that objective knowledge is

---

[2]Contemporary logic was an exercise in philosophy of science and textbooks in logic
routinely included a chapter dedicated to probability theory. For the most part they
favoured a subjective interpretation of probability. See for example the textbooks on
logic by Sigwart [57] and Lotze [36]. The alternative notion of an objective possibility
was emphasised by Lange [34] and Adolf Fick [14]. Thus, the work of von Kries was
situated in a very active research field.

inevitably necessary to answer the question of what constitute cases of equal possibility:

> It has to be particularly emphasized that a mere lack of knowledge regarding a specific arrangement does not warrant the assumption of a particular probability statement. [67, p. 59]

2.3   A game of chance

Before examining the conditions of measurement developed by von Kries, an illustrative game of chance is first be considered:

Von Kries considers a game he called *Stoßspiel*.[3] In principle, the game is a simplified form of roulette and works as follows ([67, pp. 49ff.]; compare also [28, 29]):

> There is a small groove and a ball. The groove is marked with thin black and white stripes; ideally their limiting width converges to zero. The ball is forced to start rolling along the groove. These are the conditions determining the behaviour of the ball. The frictional loss means that at some point the ball will come to a halt. Dependent on the intensity of the force initially given to the ball there is a certain range for the stopping point.

Von Kries poses the question: will the ball stop on a black or a white stripe? He asks for the size of the range left open for black and white. This is what is meant by the term *Spielraum*. Of course, even though the probabilities that the ball will stop on the 50th or 2000th stripe are quite different, the overall range for white equals the range for black. The reason for this is that the black and white stripes are very thin and a black stripe follows a white stripe, and vice versa. The crucial point is, however, that the probability for the ball to stop on, say, the 1000th stripe, which might be white, or on the 1001th stripe, which would then be black, is nearly the same. Since this holds for every two stripes under consideration, the overall probability for black and white is the same, namely $\frac{1}{2}$; both regions of possible behaviour are of equal size.[4] It is of particular interest that we consider the mechanism

---

[3]A modern and very similar analysis is given by Suppes [62, 63], who investigates the mechanics of flipping a coin. With far less mathematical expenditure, but by and large producing the same result, this example was already introduced by Fick [14]. In an overview of the history of probability theory von Kries claimed to have provided a more precise formulation of Fick's account [67, pp. 266ff.].

[4]At the beginning of the 20th century the Polish physicist Marian Smoluchowski developed a model to explain laws of chance [74] that also shows some similarities to this game of chance [70]. Starting from a classical game of chance such as throwing a dice, he developed a "purely physical model of 'regular' chance" [74, p. 262], which obeys the principles of games of chance: namely, that small variations in starting conditions produces chance. Consider a container with a small hole and a ball moving inside this

applied to create chance that is, small variations in the starting or boundary conditions, respectively.[5]

2.4   Conditions for measuring single case probabilities

By analysing this example, von Kries developed theoretical conditions that a concrete determination of probability values has to fulfil. Within the context of scientific reasoning, the most straightforward method of obtaining objective knowledge is measurement. Hence he argued that the probability values should to be measured in some way. To achieve this goal, he developed measure theoretical conditions. We should pause to note that von Kries was a professional physiologist. A field of research field that, especially in Germany [76], was primarily responsible for the emancipation of psychology from philosophy [52]. Physiological laboratories were founded at many universities [75]. Von Kries worked in the field of the physiology of the senses. For instance, he published the third edition of Helmholtz's Physiological Optics, and made significant contributions to the investigation of eye perception. He therefore was obliged to deal with problems of measurement professionally. Hence, while the philosophical approach to probability through the theory of logical ignorance relied on the method of introspection, by contrast his frame of reference was the scientific practice established in the 19th century, that aimed at determining objective facts using a process of scientifically controlled measurement. It is thus the measuring device itself that guarantees the objectivity of probability values.

The work of contemporary physiologists was devoted to the measurement of so-called intensive qualities. The notion of intensive qualities was a technical terminus in contemporary physiology and psychology: it refers to mental sensations. With the beginning of psychological measurement, the issue of if and how psychological properties could be measured was contested, since these properties cannot be directly observed in the same way as physical entities. The difference was denoted by the dichotomy of exten-

---

container. The probability that the ball will be thrown out of the container can be calculated as a function of the speed of the ball, the volume of the container and size of the hole. The probability is thus

$$W = \frac{\omega c t}{4V}$$

With ω as the size of the hole, c the velocity of the ball and V the volume of the container. The mean time, in which the ball stays inside the container can now be calculated:

$$T = \frac{4V}{\omega c}$$

This can be interpreted as the half-life in the case of radioactive decay. Thus the idea that this represents a *purely physical* model of chance indicates that Smoluchowski was still searching for a mechanistic explanation of radioactive decay.

[5]Within the context of differential equations, this forms the starting point for the modern theory of determinstic chaos.

sive and intensive qualities: for instance, pain, fear or the subjective feeling
of heat are intensive qualities, but also degrees of expectation in terms
of tossing heads or tails. As early as 1882, von Kries posed the question
that addressed theoretical conditions for measuring psychological qualities
in a paper on 'the measurement of intensive qualities and the so-called psy-
chophysical law'. In this paper, he investigated the theoretical conditions
that every measurement has to fulfil; this study was undertaken a whole
five years before Helmholtz's famous Paper on 'measuring and counting'.[6]
Thus, von Kries counts as one of the earliest contributors to the theory
of measurement [38]. Contrary to modern insights he gave equality logical
priority over 'greater or equal' ordering. Today logical priority is given to
the relation 'greater or equal', $\geq$, over the equality sign, $=$, since equality
can be defined as $A \geq B$ and $B > A$. Von Kries, however, claimed that "if
the meaning of equality is determined then it is also obvious what is meant
by the multiple" [66, p. 259]. In fact, it is possible to build up an ordering
structure from this starting point [30, pp. 15ff.]. To reach this goal, it first
has to be determined what, precisely, it means to say that one feature of
two entities is of the same size. Von Kries realised that every measurement
has to fulfil the condition that the quantities on the left and right side of an
equality sign are of the same dimension. This is obvious in the case of the
comparison of, for example, the length of two rods. Von Kries generalised
this insight in the statement that every measurement has to be reduced to
the comparison of space and time; that is, every measurement has to be
reduced to extensive quantities. Note, however, that the theory of measurement referred to by modern textbooks [30, 43] did not emerge until the
1950s.

2.4.1 Extensive measurement

Von Kries then applied this measure theoretical insight to the problem of
determining cases of equal possibility. In modern terms, it could be said
that what he needed was an extensive measurement structure. Note that
an extensive measurement determines objectively given measurement values. To put it in contemporary terms: to exist as a basic primitive concept,
and not in terms of probabilities, *Spielräume* have to represent an empir-
ical relative $\langle A, \succeq, \circ \rangle$ for the numerical relative of probabilities. The most
prominent example for an empirical relative in the measure theoretic liter-
ature [43, pp. 47ff.] is the comparison of the length of two rods. The rods
are the set of a physical entity A, and their relative length is the empirical
counterpart of the respective numerical relation $\geq$. If we folded together
the two rods, we would have the empirical relative of the mathematical op-

---

[6]It has to be noted, however, that he worked in 1876 at Helmholtz Laboratory in
Berlin, and it is very likely that he heard Helmholtz's lectures on measuring.

eration of addition. In our case, A is the symbol for a *Spielraum*, $\succeq$ is a binary relation and $\circ$ a binary operation. However, von Kries did not say anything about the operation, which is in any case given by classical probability theory. In the technical part of his theory, he developed a measure theoretic framework for the binary relation. The meaning of the *Spielraum'* A will be considered in the next section. As a precondition for extensive measurement this relation has to fulfil the axioms for a weak ordering [43, p. 47]. Formally expressed, the relation $\succeq$ has to satisfy

1. if $a \succeq b$ and $b \succeq c$ then $a \succeq c$ (Transitivity), and

2. $a \succeq b$ or $b \succeq a$ (Connectedness).

   Von Kries developed three conditions for the comparison of the size of two *Spielräume*: they have to be indifferent, original, and comparable. In the following it will be argued that these conditions are sufficient to establish a weak ordering; that is, a relation that is transitive and connected.

(1) Indifference. The first condition is developed as a means of introducing the relation and is called indifference. Indifference is the first condition that is obviously necessary for equality. Since von Kries is concerned with the determination of *equally* possible cases, he develops the ordering structure not through the asymmetric relation $\succeq$ but rather by the symmetric relation $\sim$.

   The relation of two parts of a *Spielraum* is termed indifferent if a whole namely, a *Spielraum* exists, that can be subdivided into equal units[7]. For instance, a regular die consists of six sides of equal size. For another example, recall the example of the '*Stoßspiel*': it is a feature of the arrangement of this game that the black and white stripes are of equal size. Thus the groove can be subdivided into black and white parts. In order for this subdivision to be indifferent, there has to be some kind of knowledge that there are no reasons in favour of black or white. If there were for example, a magnetic field connected with the black stripes, and the ball was made of iron, then there would be reasons in favour of black. The player has to know that this is not the case. "Only under this precondition is the logical legitimacy of singular assumptions exclusively dependent on the size of the area they comprise" [67, p. 25]. The expectation is only based on the size of the field of possible behaviour and not on any causal considerations. In this case, two units A (say, the black stripes) and B (say, the white stripes) under consideration fulfil the relation $\sim$; i.e., the scope of both stripes is of equal size: $A \sim B$.

---

[7]Note that the whole defines the maximum probability, $p = 1$. Thus there is no analogy to the archimedian axiom which states that rods can be lined up to infinity.

At first glance, this looks very much like the core principle of subjective probability interpretations, the principle of insufficient reason. This states that two cases are of equal possibility if nothing is known in favour of either one or the other. But it has to be remarked that von Kries insists that one has to *know* that there is no reason in favour of one of the alternatives. Hence, this reasoning is not based on a lack of knowledge, but rather on positive grounds. This scenario is exemplified by the '*Stoßspiel*'. In this game, the degree of expectation for throwing white or black respectively is only dependent on the size of the black and white stripes. That the probability value for white is $\frac{1}{2}$ is due to the fact that the range of white stripes comprises half of the total. With the exception of the starting conditions, all other conditions of the game are precisely known. For instance, it has to be known that there are no holes in the groove. Irrespective of the intensity of the concrete strength of the force given to the ball (assuming the force remains within a certain range and will not, e.g., be such that the ball is thrown out of the groove), measuring the probabilities for black and white is reduced to the comparison of the sizes of the black and white fields. One therefore knows that one does not need to know the starting conditions for determining the probability values[8]. In particular, applications of the probability calculus in the field of inductive reasoning are thereby prevented.[9] To vary a statement already made by Kurt Grelling [20, p. 464], it is not a principle of insufficient but rather of sufficient reason.

(2) Originality. Furthermore the condition of originality has to be satisfied. This is a condition that establishes the transitivity of the relation $\sim$. The comparison of the size of two ranges must not lead to different results at different times. Thus, the size of a *Spielraum* (and hence the probabilities measured by the *Spielraum*) cannot change in time; that is at any given time $t_0 < t_1$ the result of a comparison would have to be the same. In terms of the theory of measurement this condition implicitly guarantees the transitivity of the relation $\sim$. If this condition is not fulfilled, intransitivity would be possible. At time t_1 there might exist a relation $A \leq B$ and $B \leq C$ and hence $A \leq C$. However, if the condition of originality is not fulfilled, it might be possible that at t_0 the relation could take the form $A \leq B$ but $B \geq C$. In this case, it might be that $A \geq C$. This is a crucial point, since von Kries was aware that intensive quantities in psychological measurement are quite often intransitive. Comparable objections hold for many cases that might look like instances of probabilities.

Again this condition can be illustrated by the '*Stoßspiel*'. The momentum given to the ball is dependent on the force and the motion of the hand that rolls the ball along the groove. It depends on numerous arbitrary con-

---

[8]If the starting conditions were known, the probability would be 0 or 1.

[9]This interpretation attracted attention in France; in particular [44, 51].

ditions, which strictly speaking include the complete history of the event
[67, pp. 70ff.]. However, this does not change the probability for getting a
white stripe. The considerations that lead to the degree of expectation are
of such a general kind that the result will not change when the history of
the event is also taken into consideration. It does not matter if the force
is comparatively low (so that the ball stops, say, at the 50th stripe) or if it
is comparably strong (so that the ball stops at the 500th stripe). Historical considerations about the strength of the initial force are irrelevant for
determining the probability value.

(3) Comparability. Finally, the connectedness of the relation has to be
guaranteed. Therefore two quantities have to be compared in an unequivocal manner. Von Kries characterised the condition of comparability as
a special kind of measuring condition [67, p. 35] For example, the process
of throwing a dice (and, of course, the process of throwing a ball along a
groove) is composed out of an overwhelming mass of different factors such
as the contraction of muscles, the direction of the impulse given to the die
(or the ball) and many others. All these factors can by no means be determined numerically. No scale unit exists to compare two throws of a die.
Nevertheless, the range of possible behaviour has a clearly fixed numerical
value [67, pp. 55, 69].

The way he solved the problem to guarantee an unequivocal comparison
was to compare ranges within the smallest neighbouring units ([67, p. 64],
cf. [28, p. 246]).[10] For example, the property of a farmer may consist of
sheep, cows, fields, and so on. It may also be assumed that the monetary
value (this would be a rule) of this property is unknown. It is nevertheless
possible to determine what half of the property is by splitting every single
item into two parts. Thus, if for example the total property consists of
10 sheep, 20 cows, and so on, half of it would consist of 5 sheep, 10 cows,
and so on [67, p. 66]. Without knowing the monetary value of the whole
property, single items like sheep, cows etc. can be divided one by one into
two halves. In this manner it is possible to compare two ranges without a
common third.[11] In case of a *Spielraum*, this is possible if two conditions
are fulfilled.

1. The space of possible outcomes has to be strongly intermixed [29,
   p. 318]: if two possible events are compared then the regions in which
   the one event will happen and the regions in which the other will

---

[10]This has been pointed out as a possibility for retaining a non-modal substitute for
modal chances in the face of counter examples [37, p. 60].

[11]Contemporaries did not find this easy to see. Adolf Elsas [12, p. 566] criticised the
argument by claiming that *Spielräume* could only be measured if there was a rule as a
common third and consequently also rejected the idea that a property can be divided
without having a monetary value.

occur follow one after the other in a very short distance. Again, this
condition can be explained using the example of the *Stoßspiel*. The
space of possible outcomes is given here by the values black or white.
However, a thin white stripe follows a black and vice versa; the space
of possible outcomes is thus strongly intermixed.

2. A certain function for the concrete outcome has to be defined over this
   highly abstract space of possible outcomes. To return to our example
   of the *Stoßspiel*, this function would be the one to calculate the point
   at which the ball will come to a halt. It has to describe the motion
   of the hand throwing the ball; it would be impossible to describe the
   function. However, it is not necessary to know its concrete form, but
   only to know that it has a certain characteristic: namely, "slightly
   different values of the argument have to result in only slightly different values of the function" [67, p. 51]. This is clearly the case in
   our example. This is exactly the definition of strong causality [56]:
   in particular chaos theory commonly distinguishes between weak and
   strong causality. The principle of weak causality only demands that
   the *same* arguments result in the same outcome, but says nothing
   about the outcome of similar arguments. As chaos theory demonstrates, it could be the case that slight variations of the argument
   might result in completely different outcomes. On the other hand,
   the principle of strong causality demands that slightly different values
   of the argument have to result in only slightly different values of the
   function. Systems obeying the principle of strong causality are linear
   systems. Strong causality is a stronger criterion than weak causality
   since it is precluded that small variations of the input lead to completely different output values. For instance, feedback control is only
   possible under the condition of strong causality. Thus, von Kries'
   considerations point to the principle of strong causality.

The crucial point is that the function in condition (2) is defined over the
strongly intermixed space of possible outcomes in condition (1). Therefore
von Kries is able to derive the comparability within smallest neighbouring
units. This can again be illustrated again by the *Stoßspiel*: A slight variation in the momentum given to the ball will result in a slight change in the
distance covered by the ball. This is due to the fact that the behaviour of
the ball obeys the principle of strong causality of condition (2). However,
because a thin black stripe is followed by a white stripe, that is, the space of
possible outcomes is strongly intermixed in the sense of condition (1), this
small change will change the result into either black or white. Since this
consideration holds for every two stripes, the overall probability for black
or white respectively is $\frac{1}{2}$.

If the system under consideration fulfils these two conditions, it is guaranteed that two entities can be compared. However, if they can be compared then it follows that one of the two is greater or equal than the other, that is, the condition of connectedness is fulfilled.

2.5   First result: A measurement structure for single case possibilities

The results obtained so far can now be summarised as follows: With the notion of indifference von Kries has established a relation $\sim$. This allows us to identify cases of equal possibility. With the notion of originality and comparability he has provided criteria for the transitivity and connectedness of this relation. Therefore he has solved the problem of establishing a weak ordering structure. Note that these conditions have to be fulfilled in every single instance of a measurement. In fact, von Kries claimed that the explanation of relative frequencies have to be gained by a reduction of the 'law of large numbers' to single cases [67, p. 89, pp. 159ff.].[12] Thus, contrary to modern debates on propensities, the measurement of single case probabilities is not gained by speculation or by counting relative frequencies. Note that *Spielräume* are an objectively given empirical relative, since their determination is a result of a measurement process. Therefore von Kries' approach can be subsumed under the tradition of the scientific movement of the 19th century [52, 59]. Once cases of equal possibility are determined, probability values (for finite sets) can be determined according to the Laplacian definition; namely by counting possible and favourable cases.[13] On this basis, then, it is possible to determine favourable cases.

Yet the question remains whether such a highly demanding measure theoretical structure can be applied to any processes in nature. Von Kries argued that this is indeed possible in the case of Ludwig Boltzmann's probabilistic explanation of the approach to equilibrium in thermodynamics processes. Von Kries' argumentation will be investigated in the next section.[14] It turns out that it also sheds light on the relation of the *Spielraum* theory to the question of determinism.

---

[12]In the 19th century the term 'law of large numbers' denoted not only a mathematical theorem but also a fact of empirical nature: the existence of relative frequencies, as is it nicely illustrated by Poisson [44].

[13]In this way, von Kries felt he had elaborated his introductory example of a meteorite hitting the earth. It is in principle possible to divide the surface of the earth in an unequivocal way. Yet —something that is *not* considered by von Kries— this is highly demanding: to take into account that the earth is a globe and not a disk, one has to know the direction from which the meteorite arrives. To take into account rotation, even the time of the crash has to be known. According to his conditions, the division of the surface of the earth would not be original. This example demonstrates how demanding these conditions actually are.

[14]Since the development of physics has gone beyond Boltzmann, obviously this can only be understood as an historical argument.

3   Second part: Objective single case probabilities and determinism

As a second step, the relation between single case probabilities and determinism will be analysed in further detail. The game of chance developed by von Kries already suggests that the *Spielraum* theory of probability makes no contribution to an indeterministic world view. We should note that von Kries developed his theory with respect to some recent developments in the research programme of kinetic theory of gases: namely, the introduction of probabilistic reasoning in Ludwig Boltzmann's work. In fact, Boltzmann regarded von Kries' theory as the logical foundation of his own utilisation of probability theory [4]. Even though von Kries later realised that his *Spielraum* theory can be applied also in very different problem fields [68], in his 1886 book von Kries relied manly on Boltzmann's considerations. A brief sketch of the relation of determinism and probability within the framework of the kinetic theory of gases will therefore be given, before von Kries' considerations on probabilistic reasoning within a deterministic framework are examined in more detail. This will explain the meaning of the *Spielraum* and hence, establish a categorial basis for probabilistic reasoning.

3.1   Thermodynamics and determinism

Briefly speaking, the research programme of the kinetic theory of gases was to reduce the phenomenological laws of thermodynamics to the laws of mechanics. The idea that heat could be explained by the motion of molecules can be traced back to 1738 [6]. Until the beginning of the 19th century, competing theories were of equal credibility [6, 7]. However, in the 19th century Newtonian mechanics was the very ideal of a scientific explanation [64, 32]. Yet thermodynamics remained a challenge to the explanatory power of science [8]. To gain a mechanical understanding of thermodynamics, a gas was modelled as a set of very many identical particles. These hypothetical atomic elements where thought to have no inner structure and to move in a chaotic manner through the gas container. This is all what is needed for the model; no assumptions about their moving trajectories have to be made. However, it is clear that they will collide and that this collision will follow mechanical laws of elastic collision [8]. By colliding, 'atoms' exchange energy. In this respect, Rudolf Clausius spoke of the motion we call heat [9]. The kinetic theory of gases thus is in accordance with classical mechanics.

In the case of systems approaching equilibrium, however, this programme proved to be problematic [26, 2, 65]. Note, that the approach to equilibrium is an irreversible process,[15] whereas the laws of mechanics are reversible. It

---

[15]In contemporary debates this was related to the 2nd law of thermodynamics [64], commonly attributed to Clausius [5]. In fact, many different formulations of the 2nd law exist [65, 5]. One formulation states that in a closed system entropy cannot decrease.

was thus argued by J. J. Loschmidt [35] that an explanation of the process
of a system approaching equilibrium cannot be reduced to the laws of mechanics. Ludwig Boltzmann struggled with probabilistic arguments in an
effort to counter this argument [3]. In particular, he investigated the starting conditions of the atomic particles: if nothing is known about the starting
conditions, then every starting condition is of equal possibility. Boltzmann
then calculated in a purely combinatorial way how many of these conditions
would cause an unequal distribution of heat and how many would cause an
equal distribution of heat. At this point, probability theory comes into play:
namely, the result was that the overwhelming majority of the starting conditions would result in the latter scenario. It is of overwhelming probability.
This means that no purely mechanical explanation of the approach to equilibrium is possible, since the starting conditions also have to be taken into
regard. A kinetic explanation is an explanation via the universal (that is,
timeless) laws of mechanics *plus* the starting conditions. Nevertheless, the
kinetic theory of gases does not rely on an indeterministic generalisation of
the concept of causality. All that is needed are the laws of classical mechanics. These laws, however, are of a deterministic nature. The application of
probability theory is restricted to the starting conditions [58, 73]. It should
be noted that this idea is covered by the games of chance described above.

3.2   Von Kries on the second law of thermodynamics

In his book on probability theory, von Kries attempted to demonstrate that
Boltzmann's considerations fulfil the conditions of indifference, originality
and comparability. In fact, Boltzmann was content with von Kries' explanation of the nature of his probabilistic arguments [4]. Obviously, since
1886 there has been a long-lasting debate on this topic [58]. In particular,
it has been demonstrated that Boltzmann's account failed. However, this
failure is due to Boltzmann's assumption of ergodicity, which was relaxed
by von Kries, as we will see below. Nevertheless, von Kries' account is not
a contribution to the current debate in physics. It is an outline of how his
measure theoretic considerations can be applied, and hence of what von
Kries considered as to be a correct application of probability theory. This
allows for specifying the object of probabilistic explanations.

Without going into details [65], this is closely connected to the observation that heat
cannot be transferred from a body with a lower temperature to a body with a higher
temperature. An examples is the mixture of two different gases. While it is a common
notion that the 2nd law is about irreversibility [58], it has been objected that this actually is untrue [5]. It all depends on the formulation. It is argued that the problem
of irreversibility might be better explained by the so-called zero law of thermodynamics introduced in the 20th century [5]. For the purposes of investigating the object of
Spielraum theory it is sufficient to note that irreversible processes are a problem for
mechanistic explanations, without going into the details of this long-lasting debate.

(1) Indifference. Von Kries took it for granted that the condition of in-difference is fulfilled simply because the central assumption of the model is that nothing is known about the inner structure of gas. At this point it becomes evident that von Kries' theory is an historical account: von Kries did not pay much attention to the problem, and in particular not to its history, that there are many different possibilities in constructing a model. For instance, in an early paper, Krönig [31] simply differentiated between six groups of molecules with equal velocity and vertical direction of motion. For reasons of simplicity, further contrafactual assumptions are introduced in this model. Nevertheless, one would be in a state of indifference between these six possibilities. In the work of Clausius, Maxwell and others (in par-ticular Boltzmann), this simple assumption was replaced by increasingly realistic assumptions [58]. There are numerous possibilities for the model construction. Indifference is thus not sufficient to justify any particular mi-crocanonical measure. However, indifference is a precondition for von Kries' theory of measurement. In [67, pp. 39ff.], he discusses a 'fictitious example' of an idealised material particle: nothing is known about it except that it is within a space V. In this case, it is equally likely that it is within any certain range v of this space V. The probability that it is in any of these parts would be $\frac{v}{V}$. Similar considerations hold for the velocity and direction of the motion of this particle. Von Kries argued that a further elaboration of these considerations can be applied to the kinetic theory of gases.[16] If nothing is known about the inner structure of the gas, then the *Spielräume* of the behaviour of this particle are indifferent [67, p. 208]. However, it has to be guaranteed that the situation will not change if the causal history of the motion of this particle is taken into consideration [67, p. 67]. This leads to the condition of originality.

(2) Originality. Von Kries argued that the condition of originality was fulfilled; in principle, this amounts to the so-called Liouville Theorem. The theorem proves that every factual distribution of velocities can be reduced to a prior distribution of velocities consistent with the assumption of equal possibility of every velocity [27]. However, von Kries struggles with the physical mode of speech which states that the state of a system changes from an improbable state into a more probable one. This mode of speech suggests that the probabilities appear to change over the course of time. In that case, the condition of originality would not fulfilled. Yet, as von Kries points out [67, p. 200], this mode of speech does not denote the size of a *Spielraum* for a *single* molecule. In fact, Boltzmann's assumptions are so construed that the most probable starting conditions are those that change the actual state of the system from the so-called improbable to the so-called probable one. It is, therefore, not correct to say that the probability of the

---

[16]For an example of such considerations, compare footnote 4.

state of system changes at all. The size of a *Spielraum* of one single molecule does not change in the course of time ([67, p. 200]; [69, p. 628]). It is exactly this condition that is responsible for the behaviour of the system: that is to say that the starting conditions, i.e., the range of the *Spielraum* of the molecules, results in an equilibrium. Therefore, von Kries concludes, the condition of originality is also fulfilled.

(3) Comparability. It has to be proven that the determination of the distribution of velocity of the molecules by Maxwell distribution is unequivocal. The solution suggested by Ludwig Boltzmann has become known as the ergodic thesis which has, in fact, failed. According to the ergodic thesis, it has to be proven that over the course of time a thermodynamical system like a gas will realise every possible behavioural state. This means that a trajectory of a molecule reaches every point within the phase space. It has been shown that this thesis cannot be proven mathematically [58]. Therefore this hypothesis was later replaced by P. and T. Ehrenfest with the less demanding quasi-ergodic hypothesis, which merely states that the trajectory comes *near* to every point in the phase space with any certainty (however, compare [72]).

In terms of von Kries' theory of measurement, this means that it can be shown that the condition of comparability has been fulfilled. For the distribution to be unequivocal, it has to be proven that a gas cannot "realise several completely different modes of behaviour, A, B, C, which would last permanently, so that if A takes place, only a certain category of states could alternate, and if B takes place then only a certain completely different category of states, etc., but that the realisation of A permanently excludes the occurance of B" [67, p. 209]. If this were be the case, the *Spielraum* would not be connected. However, since von Kries assumes that Boltzmann's thesis holds, he assumes that the condition of comparability is fulfilled: namely, for this condition to be fulfilled it has to be guaranteed that two trajectories are comparable within the smallest neighbouring units. Since the ergodic thesis claims that a trajectory reaches *every* point in a phase space, it is a forteriori possible for two states S_1 and S_2 to find a *small range* of states $d\sigma$ in the history of the system, including a σ_1 and a σ_2 with only minimal differences, but with σ_1 leading to S_1 and σ_2 leading to S_2 [67, p. 210]. Note that von Kries only demands comparability within smallest neighbouring units. For this condition to be fulfilled it is sufficient that the trajectory comes near to every point in the phase space. Hence, in principle, von Kries' account is more in line with the quasi-ergodic thesis (developed nearly 30 years later) than with Boltzmann's original ergodic thesis.[17]

---

[17] However, von Kries examines a different model: he considers a 6-dimensional space of an individual molecule and not the $6N$-dimensional phase space of a gas.

3.3   A deterministic framework for single case probabilities

With these considerations von Kries aimed to show that the probabilistic argumentation in the kinetic theory of gases in fact fulfils the *formal* conditions to be considered as a *Spielraum*. Thus, Boltzmann's paper from 1877 can be regarded as a model of the *Spielraum* theory of probability. Obviously, this result does not pertain by accident. It is precisely the idea of investigating the influence of starting conditions that serves as the backdrop of the *Spielraum* theory of probability. However, the question remain: what is a *Spielraum* actually? That is to say, what is the meaning of probability theory?

It is worth noting, that von Kries is not concerned with the statistical properties of an ensemble of gas molecules, but instead investigates the properties of a single molecule — a crucial distinction for him. According to von Kries, it is a imprecision in the mode of speech when the state of a system is characterised by the *number* of molecules with certain properties [67, p. 203]. To understand his reconstruction of probabilistic argumentation, it might be helpful to consider the analogy and differences between Boltzmann's and von Kries' account in more detail. The analogy can be made clear using the example of the game of chance developed by von Kries. Analogously to the model of the kinetic theory of gases, the throwing of a ball along a groove can be completely explained in terms of classical mechanics. Thus, chance is created within a purely deterministic framework. Like the behaviour of the molecules the behaviour of the ball obeys only the laws of classical mechanics. But even though these laws are perfectly deterministic, the range of starting conditions produces an uncertainty in our knowledge. Under certain circumstances, however, the range of starting conditions produces exact knowledge about probabilities. The difference between Boltzmann and von Kries is that in the former's combinatorial account the range of behaviour is characterised by the number of molecules in a certain state, while the latter considers a single throw of a ball. Von Kries describes the circumstances that allow for an exact knowledge of the range of starting conditions by his measure theoretic conditions. The *Stoßspiel* demonstrates that the size of a *Spielraum* is not determined by mechanical laws but by the singular circumstances of a specific event [67, p. 84]. For this reason von Kries argued that the source of probability has to be found in these *singular* conditions of some specific event. He denoted this as the distinction between the nomological and the ontological aspects in the description of nature ([67, p. 86]; cf. [20, 24, 41, 42]). According to von Kries this is a general epistemological principle ([67, p. 87]; [69, p. 414]); even though it is the source of probabilities, it is not restricted to probabilistic arguments:

> The cognition of reality is a task that can be split into two parts.
> ... For example, the knowledge of the law of gravitation teaches us

nothing about the factual motion of the planets. We also have to know
which masses really exist and in which state of division in space and
motion they had been some time ago. [67, p. 85]

In this example, the nomological aspect is to be found in the law of gravita-
tion, whereas the existence and distribution of masses represent the ontological aspects. Uncertainty in the ontological part of the description thus says
nothing about the deterministic or indeterministic character of the nomological part. They represent kinds of knowledge which are independent of
one another. Following von Kries, probabilistic reasoning is restricted to the
ontological side. The point is made clearly by the example of Boltzmann's
probabilistic argumentation: nothing could be said about *nomological* determinism through this kind of probabilistic statements.[18] By developing
the theory of *Spielräume* von Kries thus found a way of reconciling the
perfect certainty of the laws of mechanics with probabilistic arguments. In
reality, we are always confronted with a certain degree of uncertainty. According to von Kries, the reason is dependent on ontological aspects of the
description of nature. As is the case in the '*Stoßspiel*' they may leave open a
certain range, that is, a *Spielraum* of possible behaviour. This state of affair
pertains even if the nomological laws are deterministic. In some cases, the
range can even be measured, namely in those cases when it can be shown
that the measure theoretic conditions of indifference, originality and comparability are fulfilled. For instance, in the case of the kinetic theory of gases,
the degree of uncertainty is exactly determined by the model assumptions.

3.4 Second result: objective single case probabilities and determinism

With the example of the kinetic theory of gases, von Kries aimed to demonstrate that the conditions of indifference, originality and comparability are
not purely theoretical, but can help to clarify the nature of probabilistic reasoning. The kinetic theory of gases served him as a model for the *Spielraum*
theory of probability; the notion of a *Spielraum* is intended as a generalisation of the notion of a phase space. The measure theoretic conditions may
control and serve as advice in the construction of a model.

However, it should be emphasised that the debate on thermodynamics
has not reached a final conclusion until now. Obviously, von Kries' theory cannot resolve the problems that still occur in the research programme
associated with kinetic theory. Von Kries' examination of Boltzmann's account is an *example* for the style of probabilistic reasoning invented by von
Kries. *Spielraum* theory provides a basis for explaining the occurrence of

---

[18]In fact, throughout his lifetime von Kries was a strong adherent of mechanical determinism. For instance, he strongly expressed this conviction four years before his death
in an article on Kant and natural sciences [71].

probabilities within a deterministic framework; this can be achieved through the distinction between nomological and ontological aspects in the description of nature. The laws of classical mechanics represent the nomological aspect within the kinetic theory. They are deterministic. On the other hand, any application of these laws has to take starting or boundary conditions into account. This is the ontological aspect in the description of nature. According to *Spielraum* theory, objective single case probabilities are not an indeterministic generalisation of the concept of causality: causality and *Spielräume* stand in a complementary relation. There thus exists a categorial basis F – namely, the range of starting conditions that enables probabilistic reasoning.

4 Conclusion

The theory of *Spielraum* investigated in this paper is an *historical* example for a theory of objective single case probabilities. It has not been further developed as a research programme in the 20th century. Presumably, it captures neither practical results and applications of modern statistics nor recent developments in the kinetic theory. Nevertheless, the historical example suggests that we might reconsider current intuitions about objective single case probabilities.

Contrary to the intuition that propensities would not make much sense if no assumption of an ontic indeterminism is made, the reliance on small variations in the starting conditions can encompass a categorial basis for objective single case probabilities within a deterministic world view. *Spielraum* theory is thus not a probabilistic generalisation of the concept of causation. However we should note that it is objective knowledge about the range of these starting conditions that is responsible for our reaching this conclusion. Moreover, it demonstrates that a measurement of single case probabilities is at least possible in principle — even though von Kries' measure theoretic conditions are highly demanding. One does not need to rely either on relative frequencies nor purely on speculation. Hence it proves the possibility and coherence of an approach to objective single case probabilities that contradicts modern intuitions.

Bibliography

[1] Jakob Bernoulli. *Wahrscheinlichkeitsrechnung*, volume 107 of *Ostwalds Klassiker der Exakten Wissenschaften*. Verlag Harri Deutsch, 1999. Translated by Robert Haussner.

[2] Günther Bierhalter. Die mechanischen Entropie — und Disgregationskonzepte aus dem 19. Jahrhundert: Ihre Grundlagen, ihr Versagen

und ihr Entstehungshintergrund. *Archive for the History of the Exact Sciences*, 32:17–41, 1985.

[3] Ludwig Boltzmann. Über die Beziehung zwischen dem zweiten Hauptsatz der mechanischen Wärmelehre und der Wahrscheinlichkeitsrechnung respektive den Sätzen über das Wärmegleichgewicht. *Sitzungsberichte der mathematisch-naturwissenschaftlichen Classe der Kaiserlichen Akademie der Wissenschaften*, 76:373–435, 1877.

[4] Ludwig Boltzmann. Der zweite Hauptsatz der mechanischen Wärmelehre. *Almanach der Kaiserlichen Akademie der Wissenschaften*, 36:225–259, 1886.

[5] Harvey R. Brown and Jos Uffink. The Origins of Time-Asymmetry in Thermodynamics: The Minus First Law. *Studies in History and Philosophy of Science Part B: Studies in History and Philosophy of Modern Physics*, 32(4):525–538, 2001.

[6] Stephen Brush. *Kinetische Theorie, Band 1*. Wissenschaftliche Taschenbücher. Akademie Verlag, 1970.

[7] Stephen Brush. *The Motion We Call Heat*. Elsevier, 1976.

[8] Peter Clark. Atomism versus Thermodynamics. In Colin Howson, editor, *Method and Appraisal in the Physical Sciences*, pages 41–106. Cambridge University Press, 1976.

[9] Rudolf Clausius. Über die mittlere Länge der Wege, welche bei der Molecularbewegung gasförmiger Körper von den einzelnen Molecülen zurückgelegt werden. *Annalen der Physik und Chemie*, 105:239–58, 1858.

[10] Andrew Dale, editor. *Pierre-Simon Laplace. Philosophical Essay on Probabilities*. Springer, 1995.

[11] Robert Lesley Ellis. On the Foundations of the Theory of Probabilities. In William Walton, editor, *The Mathematical and Other Writings of Robert Leslie Ellis*. Deighton, Bell, and Co., 1863.

[12] Adolf Elsas. Kritische Betrachtungen über die Wahrscheinlichkeitsrechnung. *Philosophische Monatshefte*, 25, 1889.

[13] James Fetzer. Probabilistic Explanations. *Proceedings of the Philosophy of Science Association*, 2:194–207, 1982.

[14] Adolf Fick. *Philosophischer Versuch über Wahrscheinlichkeit*. Würzburg, 1883.

[15] Jacob Friedrich Fries. *Versuch einer Kritik der Prinzipien der Wahrscheinlichkeitsrechnung*. Braunschweig, 1842.

[16] Maria Carla Galavotti. Comments on Patrick Suppes Propensity Interpretation of Probability. *Erkenntnis*, 26:359–368, 1987.

[17] Ronald Giere. Objective Single-Case Probabilities and the Foundations of Statistics. In Patrick Suppes, Leon Henkin, Grigore C. Moisil, and Athanase Joja, editors, *Logic, Methodology, and Philosophy of Science IV: Proceedings of the Fourth International Congress for Logic, Methodology and Philosophy of Science, Bucharest, 1971*, pages 467–484. Elsevier, 1973.

[18] Gerd Gigerenzer and Christa Krüger. *Das Reich des Zufalls: Wissen zwischen Wahrscheinlichkeiten, Häufigkeiten und Unschärfen*. Spektrum Akademischer Verlag, 1999.

[19] Donald Gillies. Varieties of Propensities. *British Journal for the Philosophy of Science*, 51:807–835, 2000.

[20] Kurt Grelling. Die philosophischen Grundlagen der Wahrscheinlichkeitsrechnung. *Abhandlungen der Fries'schen Schule*, 3, 1910.

[21] Ian Hacking. Equipossibility Theories of Probability. *British Journal for the Philosophy of Science*, 22:339–355, 1971.

[22] Ian Hacking. Prussian Numbers 1860–1882. In Lorenz Krüger, Gerd Gigerenzer, and Mary S. Morgan, editors, *The Probabilistic Revolution*, volume 2. MIT Press, 1987.

[23] Ian Hacking. *Taming of Chance*. Cambridge University Press, 1990.

[24] Michael Heidelberger. Origins of the Logical Theory of Probability: von Kries, Wittgenstein, Waismann. *International Studies in the Philosophy of Science*, 15:177–188, 2001.

[25] Victor Hilts. Statistics and Social Science. In Ronald Giere and Richard Westfall, editors, *Foundations of Scientific Method: The Nineteenth Century*. Indiana University Press, 1973.

[26] Ulrich Hoyer. Von Boltzmann zu Planck. *Archive for the History of Exact Sciences*, 23:47–86, 1980.

[27] Georg Joos. *Lehrbuch der theoretischen Physik*. Akademische Verlagsgesellschaft, 12th edition, 1964.

[28] Andreas Kamlah. Probability as a Quasi-Theoretical Concept. J. v. Kries Sophisticated Account after a Century. *Erkenntnis*, 19:239–251, 1983.

[29] Andreas Kamlah. What can Methodologists Learn from the History of Probability. *Erkenntnis*, 26:305–325, 1983.

[30] David Krantz, Duncan Luce, Patrick Suppes, and Amos Tversky. *Foundations of Measurement Volume I: Additive and Polynomial Representations*. Dover Publications, 1971.

[31] August Karl Kroenig. Grundzüge einer Theorie der Gase. *Annalen der Physik und Chemie*, 99, 1856.

[32] Lorenz Krüger, editor. *Thomas S. Kuhn. Die Entstehung des Neuen: Studien zur Struktur der Wissenschaftsgeschichte*. Suhrkamp, 1978.

[33] Lorenz Krüger, Gerd Gigerenzer, and Mary S. Morgan, editors. *The Probabilistic Revolution*, volume 2: Ideas in the Sciences. MIT Press, 1987.

[34] Friedrich Albert Lange. *Logische Studien. Ein Beitrag zur Neubegründung der formalen Logik und der Erkenntnisstheorie*. Iserlohn, 1877.

[35] Joseph Loschmidt. Über das Wärmegleichgewicht eines Systems von Körpern mit Rücksicht auf die Schwere. *Sitzungsberichte der mathematisch-naturwissenschaftlichen Classe der Kaiserlichen Akademie der Wissenschaften Wien*, 73, 1876.

[36] Hermann Lotze. *Logik*. S. Hirzel, 2nd Edition edition, 1880.

[37] David Mellor. *Probability: A Philosophical Introduction*. Routledge, 2005.

[38] Joel Michell. Psychophysics Intensive Magnitudes, and the Psychometricians' Fallacy. *Studies in the History and Philosophy of Biological and Biomedical Sciences*, 17:414–432, 2006.

[39] John Stuart Mill. *A System of Logic: Ratiocinative and Inductive*, volume 1. University Press of the Pacific, 2002. Original edition from 1843.

[40] David Miller. *Critical Rationalism: A Restatement and Defense*. Open Court, 1994.

[41] Martin Neumann. *Die Messung des Unbestimmten*. Hänsel-Hohenhausen, 2002.

[42] Martin Neumann. A Formal Bridge Between Epistemic Cultures. In Benedikt Löwe, Volker Peckhaus, and Thoralf Räsch, editors, *Foundations of the Formal Sciences IV: The history of the concept of the formal sciences*, volume 3 of *Studies in Logic*. College Publications, 2006.

[43] Bernard Orth. *Einführung in die Theorie des Messens*. Kohlhammer, 1974.

[44] Simeon Denis Poisson. *Lehrbuch der Wahrscheinlichkeitsrechnung und deren wichtigsten Anwendungen*. Meyer, 1841.

[45] Karl Popper. The Propensity Interpretation of the Calculus of Probability, and the Quantum Theory. In Stephan Körner, editor, *Observation and Interpretation: A Symposium of Philosophers and Physicists. Proceedings of the Ninth Symposium of the Colston Research Society Held in the University of Bristol*, pages 65–70. Butterworth Scientific Publications, 1957.

[46] Karl Popper. The Propensity Interpretation of Probability. *British Journal for the Philosophy of Science*, 10:25–42, 1959.

[47] Karl Popper. Quantum Mechanics without 'The Observer'. In Mario Bunge, editor, *Quantum Theory and Reality*, volume 2 of *Studies in the Foundations, Methodology and Philosophy of Science*. Springer-Verlag, 1967.

[48] Karl Popper. *A World of Propensities*. Thoemmes Press, 1990.

[49] Theodor Porter. Lawless Society: Social Science and the Reinterpretation of Statistics in Germany 1850-1880. In Lorenz Krüger, Gerd Gigerenzer, and Mary S. Morgan, editors, *The Probabilistic Revolution. Volume 2: Ideas in the Sciences*. MIT Press, 1987.

[50] Theodor Porter. *The Rise of Statistical Thinking: 1820–1900*. Princeton University Press, 1988.

[51] Adolph Quetelet. *Letters placed to H. R. H. the Grand Duke of Saxe Coburg and Gotha on the Theory of Probability*. C. & E. Layton, 1981. Reprint of 1849 edition.

[52] Gerhard Rudolph. Das Mechanismusproblem der Physiologie des 19. Jahrhunderts. *Berichte zur Wissenschaftsgeschichte*, 6:7–28, 1983.

[53] Gustav Rümelin. Über Gesetze der Geschichte. In Gustav Rümelin, editor, *Reden und Aufsätze — neue Folge*. Freiburg, i. Br., 1881.

[54] Wesley Salmon. *Scientific Explanation and the Causal Structure of the World*. Princeton University Press, 1984.

[55] Ivo Schneider, editor. *Die Entwicklung der Wahrscheinlichkeitstheorie von ihren Anfängen bis 1933: Einführungen und Texte*. Wissenschaftliche Buchgesellschaft, 1988.

[56] Walter Seifritz. *Wachstum, Rückkopplung und Chaos*. Hanser Verlag, 1987.

[57] Christoph Sigwart. *Logik*. Mohr, 1873.

[58] Lawrence Sklar. *Physics and Chance*. Cambridge University Press, 1993.

[59] Susanne Speckenbach. *Wissenschaft und Weltanschauung*. Hempen Verlag, 1999.

[60] Stephen Stigler. *History of Statistics*. Cambridge University Press, 1986.

[61] Carl Stumpf. Über den Begriff der mathematischen Wahrscheinlichkeit. *Sitzungsberichte der philosophisch-philologischen Classe der Königlich Preußischen Akademie der Wissenschaften*, 20:37–120, 1892.

[62] Partick Suppes. New Foundations of Objective Probability: Axioms for Propensities. In Patrick Suppes, Leon Henkin, Grigore C. Moisil, and Athanase Joja, editors, *Logic, Methodology, and Philosophy of Science IV: Proceedings of the Fourth International Congress for Logic, Methodology and Philosophy of Science, Bucharest, 1971*, pages 515–529. American Elsevier, 1973.

[63] Patrick Suppes. Propensity Representations of Probability. *Erkenntnis*, 26:335–358, 1987.

[64] Coloman Szily. Das Hamilton'sche Princip und der zweite Hauptsatz der mechanischen Wärmetheorie. *Poggendorfer Annalen*, 145, 1872.

[65] Jos Uffink. Bluff Your Way in the Second Law of Thermodynamics. *Studies in the History and Philosophy of Modern Physics*, 32:305–394, 2001.

[66] Johannes von Kries. Über die Messung intensiver Grössen und das sogenannte psychophysische Gesetz. *Vierteljahrsschrift für wissenschaftliche Philosophie*, 6/3, 1882.

[67] Johannes von Kries. *Principien der Wahrscheinlichkeitsrechnung.* Mohr, 1886.

[68] Johannes von Kries. Über den Begriff der objectiven Möglichkeit und einige Anwendungen desselben. *Vierteljahrsschrift für wissenschaftliche Philosophie*, 12, 1888.

[69] Johannes von Kries. *Logik: Grundzüge einer kritischen und formalen Urteilslehre.* Mohr, 1916.

[70] Johannes von Kries. Über Wahrscheinlichkeitsrechnung und ihre Anwendungen in der Physik. *Die Naturwissenschaften*, 7, 1919.

[71] Johannes von Kries. *Immanuel Kant und seine Bedeutung für die Naturforschung der Gegenwart.* Springer, 1924.

[72] Jan von Plato. Boltzmann's Ergodic Hypothesis. *Archive for the History of Exact Sciences*, 42:71–89, 1991.

[73] Jan von Plato. *Creating Modern Probability.* Cambridge University Press, 1994.

[74] Marian von Smoluchowski. Über den Begriff des Zufalls und den Ursprung der Wahrscheinlichkeit. *Die Naturwissenschaften*, 6:253–263, 1918.

[75] Willhelm Wundt. *Essays.* W. Engelmann, 1906. Second edition.

[76] Avraham Zloczower. Konjunktur in der Forschung. In Frank Pfetsch and Avraham Zloczower, editors, *Innovation und Widerstände in der Wissenschaft.* Bertelsmann Universitätsverlag, 1973.

www.ingramcontent.com/pod-product-compliance
Lightning Source LLC
Chambersburg PA
CBHW052140070326
40690CB00047B/1199